Introduction
to the
National Electrical Code®

Introduction
to the
National Electrical Code®

Paul Rosenberg

Delmar Publishers Inc.®

NOTICE TO THE READER

Cover design: design M design W

Delmar Staff
Senior Project Editor: Christopher Chien
Production Supervisor: Wendy Troeger

For information address Delmar Publishers Inc.
3 Columbia Circle, Box 15-015
Albany, New York 12212-5015

Printed in the United States of America
Published simultaneously in Canada
by Nelson Canada,
a division of The Thomson Corporation

10   9   8   7   6   5   4   3   2   1   XX   99   98   97   96   95   94   93

Library of Congress Cataloging-in-Publication Data:

Rosenberg.
    Introduction to the National Electrical Code®/ Paul Rosenberg.
        p.   cm.
    Includes index.
    ISBN 0-8273-5305-7
        1. Electric engineering—United States—Insurance requirements.
    2. National Fire Protection Association. National Electrical Code
    (1993)   I. Title.
    TK260.R66   1993
    621.319'24'021873— dc20   93-19186
                                    CIP

Contents

PART FIVE   EQUIPMENT

PART SIX   SPECIAL LOCATIONS AND SYSTEMS

Preface

A thorough knowledge of the *National Electrical Code*® is essential to performing electrical installations of any type. The Code is a basic safety standard, and must be followed. To do otherwise would be to court disaster. Therefore, the first order of business for anyone seeking to join the electrical trade is to gain a thorough working knowledge of this Code.

Unfortunately, the *National Electrical Code*® (which we will often abbreviate as "*NEC*®," or "the code") is a long, complex, and sometimes confusing document. Learning the code can be a challenging task for newcomers to the trade.

The difficulty that beginners (and sometimes people who are not beginners) have with the code is something that has been very apparent to me, especially since I took my first job as a supervisor. Problems understanding what the code really means are epidemic in the electrical industry. And the problem is most severe, and most intimidating, to those who are just entering the trade. Many times I have tried to comfort apprentices by saying "Look, a hundred thousand other guys have learned this stuff, and you'll get it too, if you don't give up. . .". But even though this may be true, *National Electrical Code*® textbooks have been notoriously difficult to understand. (The instructors understand, but the students sit there in a daze.) Because of these difficulties, we have designed this book, INTRODUCTION TO THE NATIONAL ELECTRICAL CODE®, to be as easy-to-use as possible.

Much of the structure of this text, and many of the ideas and subjects in it, come from my own experience as an apprenticeship and Community College instructor. I had lots of firsthand experience as to what my students understood easily, and what they had difficulty with. I have tried hard to describe things in this book as if I were actually speaking to my classes. In some cases, I imagined myself standing on a job site, and explaining these subjects to a confused electrician, and tried to find the best way of explaining the subject to him. If I was successful, this textbook should be fairly easy for you to understand; clearly explained with a minimum of wording. Hopefully, it should be good enough that you could understand almost all of it even if you didn't have an instructor.

One of the first things we decided on for this book was to divide it into sections that mirror the steps that actual electrical installations follow. The first section, which covers Basic Requirements, explains the fundamental concerns of the Code, and will give you a lot of help in understanding the sometimes difficult language used in it. The second section covers raceways, which are usually the first types of equipment installed in an electrical construction project. The third section covers power distribution, or the methods and mechanics of getting power from the utility company's lines to every part of the structure being wired.

The fourth section of this book is Circuit Wiring. Circuit wiring covers all of the various types of cables, wires, and associated equipment that are used to carry electrical power to each item that uses it. Section five covers the various types of electrical equipment covered by the *NEC*®. This includes almost any type of electrical equipment that exists with requirements for each. The final section, section six, covers the code requirements for special locations (such as hospitals, airplane hangars, pumping stations, etc.) and special types of electrical systems (high voltage wiring, communications wiring, emergency systems, etc.).

The book's organization should accomplish two main things: First, it should make learning somewhat easier, since all the various subjects are grouped together according to their uses. Second, it should make a very easy-to-use reference book for later use. Almost every installation requirement that the Code contains is in this book. Very, very few were left out. You should be able to find these requirements quickly and easily. In addition, the way they are explained should be a lot easier to understand than the Code itself.

One other thing we did to make this book far easier to use was to ignore most of the design and manufacturing requirements that are specified in the *NEC*®. For example, the code gives special manufacturing requirements for almost every type of equipment or wiring it covers. These requirements are of no use whatsoever to the electrician. They are important only to the industrial engineer who is designing the product. Attempting to familiarize the electrical trainee with these requirements does little besides causing more confusion; for this reason, almost all such requirements are eliminated from this text. In the very rare instance that such requirements would matter to an installer, however, they were included.

You should find this book to have ample illustrations for the material covered. In many cases, you may want to spend some time reviewing them; very often, seeing something in diagrammatic form is more easily understood than typed words on a page. Learn to form mental pictures of the things you discuss in class. This is the way that the best electrical professionals plan their installations. If you can get a good picture in your mind, you will begin to understand how and why all the pieces fit together, and why they must be there.

One last word of advice: Always try to understand *why* things are done certain ways. This is the real factor that "separates the men from the boys" in the electrical trades. If you understand the basic reasons why things are done, you have a clear understanding of what you are doing. This puts you in control. If you just know a set of rules that you have to follow, you are never in control, and you are always unsure about your work, wondering if you forgot one of the rules. We have put a good deal of emphasis in this book on explaining the whys of the requirements. Pay close attention to these points; they will make a big difference to your future.

I wish to thank the many people who aided in the writing of this book, but first of all my editor, Mark Huth; I think this book was more difficult than either of us expected. I also want to thank my reviewers, who made this book as good as it could be: Larry Killebrew of Mid America Vocational-Technical School; John Cox of Brevard Community College; Robert Blakely of Mississippi Gulf Coast Community College; Allen R. Balling of Assabet Valley Regional Vocational High School; William Thornton of Strattford County Public Schools; Marcel Veronneau of Industrial Management and Training Institute; Bruce DeVoe of Broome Community College; and Anthony Von Egsren of Northeast Wisconsin Technical College. Finally, and perhaps most of all, I thank my students, who gave me the best possible training for writing a book such as this.

Paul Rosenberg

Delmar Publishers
Is Your Electrical Book Source!

Whether you're a beginning student or a master electrician, Delmar Publishers has the right book for you. Our complete selection of proven best-sellers and all-new titles is designed to bring you the most-up-to-date, technically-accurate information available.

DC/AC THEORY

Delmar's Standard Textbook of Electricity/ Herman

This full-color, highly illustrated book sets the new standard with its comprehensive and up-to-date content, plus complete teaching/ learning package, including: Lab-Volt and "generic" lab manuals and transparencies.
Order # 0-8273-4934-3

CODE & CODE-BASED

1993 National Electrical Code®/ NFPA

The standard for all electrical installations, the 1993 NEC® is now available from Delmar Publishers!
Order # 0-87765-383-6

Understanding the National Electrical Code®/ Holt

This easy-to-use introduction to the NEC® helps you find your way around the NEC® and understand its very technical languange. Based on the 1993 NEC®.
Order # 0-8273-5328-6

Illustrated Changes in the 1993 NEC®/ O'Riley

Illustrated explanations of Code changes and how they affect a job help you learn and apply the changes in the 1993 NEC® more effectively and efficiently!
Order # 0-8273-5304-9

Interpreting the National Electrical Code®, 3E/ Surbrook

This excellent book provides the more advanced students, journeyman electricians, and electrical inspectors with an understanding of NEC® provisions. Based on the 1993 NEC®.
Order # 0-8273-5247-6

Electrical Grounding, 3E/ O'Riley

This illustrated and easy-to-understand book will help you understand the subject of electrical grounding and Article 250 of the 1993 NEC®.
Order # 0-8273-5248-4

WIRING

Electrical Wiring - Residential, 11E/ Mullin

This best-selling book takes you through the wiring of a residence in compliance with the 1993 NEC®. Complete with working drawings for a residence.
Order # 0-8273-5095-3

Smart House Wiring/ Stauffer & Mullin

This unique book provides you with a complete explanation of the hardware and methods involved in wiring a house in accordance with the Smart House, L.P. system. Based on the same floor plans found in **Electrical Wiring - Residential, 11E**.
Order # 0-8273-5489-4

Electrical Wiring - Commercial, 8E/
Mullin & Smith

Learn to apply the 1993 NEC®as you proceed step-by-step through the wiring of a commercial building. Complete with working drawings of a commercial building.
Order # 0-8273-5093-7

Electrical Wiring - Industrial, 8E/
Smith & Herman

All-new content on hazardous locations, more detailed coverage of branch circuits, and calculating circuit sizes with the 1993 NEC®will help you learn industrial wiring essentials! Complete with industrial building plans.
Order # 0-8273-5325-1

Cables and Wiring/ AVO Multi-Amp

Your comprehensive practical guide to all types of electrical cables, this book discusses applications, storage, handling, pulling, splicing, and more!
Order # 0-8273-5460-6

Raceways and Other Wiring Methods/ Loyd

This excellent new book provides you with complete information on metallic and nonmetallic raceways and other common wiring methods used by electricians and electrical designers.
Order # 0-8273-5493-2

Illustrated Electrical Calculations/ Sanders

Your quick reference to all of the formulae and calculations electricians use, this handy book features illustrations, applications, examples, and review questions for each calculation or formula.
Order # 0-8273-5462-2

MOTOR CONTROL

Electric Motor Control, 5E/ Alerich

The standard for almost 30 years, this best-selling textbook explains how to connect electromagnetic and electric controllers.
Order # 0-8273-5250-6

Industrial Motor Control, 3E/ Herman

This excellent third edition combines a solid explanation of theory with practical instructions and information on controlling industrial motors with magnetic and solid-state controllers. Includes coverage of programmable controllers.
Order # 0-8273-5252-2

EXAM PREPARATION

Journeyman Electrician's Exam Preparation Book/ Loyd

Master Electrician's Exam Preparation Book/ Loyd

To request examination copies, call or write to:

Delmar Publishers Inc.
3 Columbia Circle
P.O Box 15015
Albany, NY 12212-5015
Phone: 1-800-347-7707 • 1-518-464-3500 • Fax: 1-518-464-0301

PART
1

BASIC REQUIREMENTS

CHAPTER
1

The Hows and Whys of the NEC®

Before you can really understand the *National Electrical Code®*, it is almost mandatory that you understand what this document is, who writes it, and what its overall purpose is. Understanding these facts will make everything you do with the Code easier, and will help you avoid costly errors.

We can begin by understanding that the *National Electrical Code®* is just one of many codes and standards published by the National Fire Protection Association (NFPA). The NFPA is a not-for-profit corporation set up many years ago by the insurance underwriting industry, who make their money by writing and publishing various codes and associated training materials.

THE CODE'S ORIGINS

The NFPA came into being around the turn of the century to regulate new uses of electricity and other types of fuel. Until then, there had been no rules regarding the installation of electrical wiring; and as the number of electrical wiring installations began to rise, so did incidents of electrically caused fires. Because of this, a number of insurance companies set up a committee to write rules for safe electrical installation. And after these rules were published, the insurance companies agreed that none of them would insure structures whose electrical wiring did not conform to the new rules. Thus, the *NEC®* became a set of rules that defined safe electrical installation. In addition, conformance to these rules became mandatory if the owner wished to have his building insured. In the course of time, most of the enforcement of these rules was taken over by local building inspectors, but the net effect remains the same: If these rules are not followed, the building cannot be used.

REVISIONS

Because of the way electrical wiring was changing, it became necessary to revise the Code every so often. Until fairly recently, this was done rather haphazardly. Currently, the Code is revised every three years, and it appears that this will be the standard method. As would seem obvious, the Code must be revised to keep up with new materials, tools, and methods that are constantly being developed.

The actual job of revising the Code is performed by twenty-one separate committees, each of which is comprised of approximately ten to fifteen persons, the majority of whom are engineers. These committees meet several times, discuss all proposed changes, accepting some and rejecting others, and rewrite (when necessary) the sections of the Code assigned to them.

Because of the many changes proposed every three years, updating the *NEC®* is no small chore. But the real difficulty is that this document must remain applicable to all types of electrical installations, leaving no "gaps." It then becomes a very lengthy document, with endless exceptions and qualifications. For this reason, the Code is frequently very confusing and difficult to interpret.

THE REASONS FOR DIFFICULTY

The engineers who write the Code are deeply concerned with technical accuracy and completeness. They generally do a very good job, but that does not ensure that their work will be easily understood by the people who must use the Code. In fact, it is the very effort to make the Code complete and correct that also makes it difficult to understand. A given article of the *NEC®* may contain information that is pertinent to engineers only, other information pertinent to manufacturers only, and still more information of use only to installers. Yet all of this information is lumped together into one brief article, and must be dissected by the reader. This is where the difficulty really lies.

BEYOND MINIMUM STANDARDS

The *NEC®* is written as a minimum standard for electrical installation, with the protection of life and property as its general purpose. It does not necessarily define the best or most efficient installation methods, merely the minimum safe standards. Very frequently, various sections must be combined creatively in order to make an efficient and inexpensive installation. Just using the first allowable way of doing something in the Code often results in a fine installation — but one that is far too expensive.

For instance, a 100-amp service run in aluminum conduit is a fine installation, but it costs far too much money for anyone ever to do such a thing. Using service-entrance cable will get the same job done, at a fraction of the cost. This is an obvious example, but it is only one of many, many such situations.

Because of this, it is doubly important to be able to use the Code well. Why should someone pay four thousand dollars for a system that someone else can install for three thousand dollars that will do the same thing? In other words, how well you can use the Code makes a lot of difference in dollars and cents, as well as in safety.

In addition, remember that there are many instances where the person who buys an electrical installation demands one that exceeds the requirements of the *NEC®*. Meeting Code is not always enough in the marketplace.

ADOPTION AS LAW

It has become a fairly standard practice in the United States for municipalities to adopt the *NEC®* as the standard for all electrical installations in their jurisdiction. By doing so, they make the *NEC®* the law in their area. That is why the Code is written with "legalese" wording, since it will become law in most areas of the United States.

It is, however, very important to remember that many municipalities amend the *NEC®* by adding their own requirements. For instance, one of my area municipalities requires that all work in their city be done according to the *NEC®*, except that all electrical services must be made with steel conduit. It is, therefore, critical that you check with local officials before performing an installation in a new area.

Another important point to remember is that the "authority having jurisdiction" (usually the local electrical inspector) has full power to interpret the Code as he or she sees fit. You should refer to *Section 90-4* of the Code and review this for yourself; it is of critical importance. The inspector has the final say, and there is very little you can do about it.

KEY WORDS

When reading and interpreting the *NEC*®, you must pay special attention to the following key words:

Shall — Any time you see the word *shall* in the *NEC*®, it means that you "must" do something in a certain way. You have no choice at all; either you do it that specific way, or you are in violation of the Code.

May — The word *may* gives you an option. You can do it the specific way that is stated, or you can do it another way; it is your choice.

Grounded conductor — This is almost always the neutral conductor. Take care not to let the word *grounded* confuse you. Grounded conductor does not refer to a green wire.

Grounding conductor — This is the green wire, more correctly called the *equipment grounding conductor* because it is used to connect equipment to ground.

In addition to these terms, there are other, less common terms that can also be confusing. Remember that the *NEC*® cannot be read casually. To make correct interpretations, you must consider each word individually. This requires extra work and more effort. There is, however, no other way to arrive at correct interpretations.

STUDY

No textbook, regardless of how good it is, can teach you how to make sense of the Code — unless you put a lot of personal effort into the process. There is a lot of material to learn, and the stakes are very high.

When I say that the stakes are high, I am referring to human life. You are probably rather new to the installation of electrical wiring, and more than likely you have never had any bad incidents with electricity. Therefore, you should be warned that there are risks involved. These risks are not great to you the installer (far more carpenters, roofers, iron workers, etc. get injured than do electricians), but rather to the people who live and work in the structures you wire.

Because of strict regulations and fairly good training, electrical accidents are fairly rare. They can, however, occur, and they can be deadly. Almost anyone who has been in this business for a number of years can tell you about deadly fires that began due to faulty wiring. As an installer, you are responsible for ensuring that the wiring you put in people's homes and workplaces is safe. Also be forewarned that the excuse of "I didn't know" won't work. If you are not sure that the installation is safe, you have no right to put it in. I am not trying to scare you, but I do want you to understand that electricity can kill, and requires experts to install. If you do not intend to put forth the effort required to become an expert, you should consider another field — one that cannot endanger people's lives.

Now that I've given you the tough facts, let me give you the other side of the coin: If hundreds of thousands of other people have done this, then you are also able to do it. The critical factor here is commitment. Any reasonably intelligent person can become expert at installing electrical wiring, provided they are committed to doing so and willing to do whatever it takes to reach that goal. Some may be a little faster, and some may be a little slower, but virtually all can get there, provided the commitment is present first.

The people in the electrical trade who are truly professionals make a steady living, and are rarely out of work; those who do not become proficient in their trade come and go. The difference between these two groups of people is not natural ability but the willingness to use the abilities they have. If you wish to apply your talents to learning how to install electrical wiring, you will gain a skill that can provide for your needs for the rest of your life. You have the ability and the opportunity — all you have to do is commit to it and see it through.

CHAPTER
2
Definitions

Article 100 of the *National Electrical Code*®is devoted to definitions of various technical terms. In this article, you will find only definitions that are deemed essential to applying the Code. You will not find definitions for every electrical term, only for those that the Code writers think are necessary to interpreting the Code. In addition, you may find that some of these definitions are rather difficult to understand. For example, an *equipment bonding jumper* is defined as "the connection between two or more portions of the equipment grounding conductor." This definition, while technically accurate, refers to the green wire that connects the metal case of a piece of equipment to a grounding system.

In this chapter, you will find the most common definitions used in the electrical field. Not all of the definitions of *Article 100* of the *NEC*®are here, although this chapter also contains many definitions that are not found in the *NEC*®, and that you are more likely to need than most of the *NEC*®definitions.

Not all of the definitions of the *NEC*®are included in *Article 100*. Many definitions that apply to specific articles are shown in that article only. For example, *Article 517* gives more than thirty definitions that apply only to health care facilities.

When reading the *NEC*® definitions, make sure that you read very carefully. They are written in a very technical way and can be easily misinterpreted if you read them quickly.

TECHNICAL TERMS

Accessible—Can be removed or exposed without damaging the building structure. Not permanently closed in by the structure or finish of the building.

Accessible, readily (readily accessible)—Can be reached quickly, without climbing over obstacles or using ladders.

Aggregate—A masonry substance that is poured into place, then sets and hardens, as concrete.

Alternate power source—This may be one or more generator sets or battery systems, if a battery system is permitted in the location. The alternate power source is intended to provide power during any interruption of the normal power supply.

Ampacity—The amount of current (measured in amperes) that a conductor can carry without overheating.

Ampere (or amp)—Unit of current measurement. The amount of current that will flow through a 1-ohm resistor when 1 volt is applied.

Appliance — Equipment installed or connected as a unit, to perform functions such as clothes or dish washing, food mixing, etc.

Approved — Acceptable to the authority having jurisdiction.

Automatic — Self-acting. Operating by its own mechanism, based on a nonpersonal stimulus.

Bonding — The permanent joining of metal parts to form an electrically conductive path.

Branch circuit — Conductors between the last overcurrent device and the outlets.

Branch circuit, appliance — A branch circuit that supplies current to one or more outlets that serve appliances.

Branch circuit, general purpose — A branch circuit that supplies outlets for lighting and power.

Branch circuit, individual — A branch circuit that supplies only one piece of equipment.

Branch circuit, multiwire — A branch circuit having two or more ungrounded circuit conductors, each having a voltage difference between them, and a grounded circuit conductor (neutral) having an equal voltage difference between it and each ungrounded conductor.

Building — A structure that is either standing alone, or cut off from other structures by firewalls.

Cabinet — A flush or surface enclosure with a frame, mat, or trim, onto which are mounted swinging doors.

Concealed — Made inaccessible by the structure or finish of the building.

Conduit body — The part of a conduit system, at the junction of two or more sections of the system, that allows access through a removable cover. Most commonly known as condulets, LBs, LLs, LRs, etc.

Continuous load — A load whose maximum current continues for three hours or more.

Controller — A device or group of devices that control (in a predetermined way) power to a piece of equipment.

Cross-sectional area — The area (in square inches or circular mils) that would be exposed by cutting a cross section of the material.

Cutout box — A surface mounting enclosure with a hinged door.

Device (also used as wiring device) — The part of an electrical system that is designed to carry, but not use, electrical energy.

Disconnecting means — A device that disconnects a group of conductors from their source of supply.

Enclosed — Surrounded by a case, housing, fence, or walls that prevent unauthorized people from contacting the equipment.

Exposed — Able to be inadvertently touched or approached.

Feeder — Circuit conductors between the service and the final branch circuit overcurrent device.

Galvanic action — A corrosive reaction that takes place when copper and zinc remain in contact.

Ground — An electrical connection (made on purpose or accidentally) between an item of equipment and the earth.

Hoistway — A shaftway or other vertical opening or space through which an elevator or dumbwaiter operates.

Identified (for use) — Recognized as suitable for a certain purpose, usually by an independent agency, such as UL.

Isolated — Not accessible unless special means of access are used.

Lamp — A light source. Reference is to a light bulb, rather than a table lamp.

Location, damp (damp location) — A partially protected location, such as under a canopy, roofed open porch, etc. Also an interior location subjected only to moderate degrees of moisture, such as a basement, barn, etc.

Location, dry (dry location) — An area not normally subjected to water or dampness.

Location, wet (wet location) — A location underground, in concrete slabs, where saturation occurs, or outdoors.

Outlet — The place in the wiring system where the current is taken to supply equipment.

Overcurrent — Too much current.

Pendant — Hanging electrical cords, to which are attached lampholders or receptacles.

Plenum — A chamber that forms part of a building's air distribution system, to which connect one or more ducts. Frequently, areas over suspended ceilings or under raised floors are used as plenums.

Phase converter — A device that derives 3-phase power from single-phase power. Used extensively in areas (often rural) where only single-phase power is available, to run 3-phase equipment.

Photovoltaic — Capable of changing light into electricity.

Radius (radii, plural) — The distance from the center of a circle to its outer edge.

Rotary converter — A type of phase converter. (See phase converter.)

Separately derived system — A system whose power is derived (or taken) from a generator, transformer, or converter.

Service — Equipment and conductors that bring electricity from the supply system to the wiring system of the building being served.

Service drop — Overhead conductors from the last pole to the building being served.

Utilization equipment — Equipment that uses electricity.

Whips — A flexible assembly, usually of THHN conductors in flexible metal conduit with fittings, usually bringing power from a lighting outlet to a lighting fixture.

CHAPTER
3

General Requirements

The general requirements for the installation of electrical wiring are given in *Article 110* of the Code. Although this section of the Code is fairly brief, it contains most of the foundational concepts that the rest of the Code is built upon.

Probably the most basic and important of the general requirements is *Section 110-12*, which states that all installations must be performed "in a neat and workman-like manner." In other words, all electrical installation requirements presuppose that the installer is concerned, informed, and mindful. Without this prerequisite, all other requirements are almost valueless. This point is of supreme importance.

Following are the individual requirements of *Article 110*, follow your Code book through *Article 110*, and you will see them all. Make sure that you understand them all; they are critically important. In my classroom experience, I have found that many students are somewhat afraid to ask questions, fearing that they will be thought of as stupid. *Please,* ask questions if you do not understand; as I have said, these installations can endanger people's lives if not done properly. Even if you were called stupid for asking a question, it is far better to be called stupid for asking a question than to be responsible for burning down a building due to ignorance. And in most cases, you will find that the person next to you did not understand either and is glad that you asked the question.

The following requirements are the basic requirements that apply to all electrical installations. They should be reviewed periodically by every electrical installer.

GENERAL REQUIREMENTS, *ARTICLE 110*

All unused openings in boxes, cabinets, etc. must be filled.

All equipment must be securely mounted. Wooden plugs in masonry are not allowed.

Conductors in manholes must be racked to provide a reasonable amount of access space.

Free circulation of air around electrical equipment (especially equipment that requires such air flow for sufficient heat removal) cannot be obstructed.

All electrical connections *must* be made with devices that are listed (and clearly marked) as suitable for their intended use. The term *listed* means acceptable by an independent testing agency, such as Underwriters' Laboratories, or a similar firm.

Several such devices are shown in Figure 3-1, and recommended torque levels for these connections are shown in Figures 3-2 through

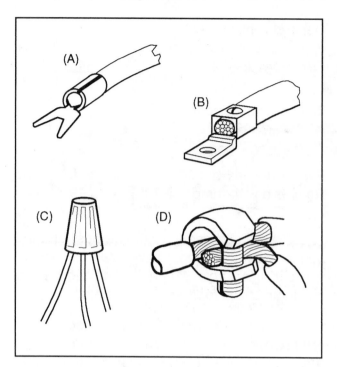

Fig. 3-1 (a) Crimp lug; (b) Barrel lug; (c) Wire nut; (d) Split bolt connector.

Tightening Torque in Pound-Feet-Screw Fit			
Wire Size, AWG	Driver	Bolt	Other
18-16	1.67	6.25	4.2
14-8	1.67	6.25	6.125
6-4	3.0	12.5	8.0
3-1	3.2	21.00	10.40
0-2/0	4.22	29	12.5
AWG 200 MCM	—	37.5	17.0
250-300	—	50.0	21.0
400	—	62.5	21.0
500	—	62.5	25.0
600-750	—	75.0	25.0
800-1000	—	83.25	33.0
1250-2000	—	83.26	42.0

Fig. 3-2 Screw tightening torque, in pound-feet.

Screws	
Screw Size, Inches Across Hex Flats	Torque, Pound-Feet
1/8	4.2
5/32	8.3
3/16	15
7/32	23.25
1/4	42

Fig. 3-3 Screw sizes and torques.

3-4. While these torque requirements are not yet part of the *NEC®*, they are quite important, since the biggest reason for failed connections is improper installation. In the majority of cases, this is due to the connections not being properly tightened. And it is not only undertightening that is problematic. Connections that are tightened too far can also fail. For this reason, the use of torque wrenches is far superior to going by "feel."

Conductors must be spliced with suitable splicing devices, or by soldering, brazing, or welding. Soldered splices must first be joined, so that the splice is not dependent on the solder for mechanical or electrical strength. All splices must be covered with insulation equivalent to that of the conductors.

A reasonable amount of working space must be provided around all electrical equipment. Generally, the minimum is 3 feet. *Table 110-16(a)* in the Code shows specific requirements.

Lighting (enough to work on the equipment) must be provided in the areas around electrical equipment.

Except for panels of 200 amps or less in dwelling units, there must be a minimum head space in front of electrical panels of 6 feet 3 inches.

Bolts			
Size	Duronze	Steel	Aluminum
Standard, Unlubricated			
3/8	20	15	16
1/2	40	25	35
5/8	70	50	50
3/4	100	90	70
Lubricated			
3/8	15	10	13
1/2	30	20	25
5/8	50	40	40
3/4	85	70	60

Fig. 3-4 Bolt tightening torques, in pound-feet.

All live parts operating at over 50 volts must be guarded against accidental contact by persons or objects. (See *Section 110-31* for over 600 volts.) The primary methods of accomplishing this are as follows:

1. By locating the equipment in a room that is accessible only to qualified persons.

2. By installing permanent and effective partitions or screens.

3. By locating the equipment on a balcony or platform that will exclude unqualified persons.

4. By locating the equipment 8 feet or more from the floor.

Guards must be installed to protect electrical equipment from physical damage where necessary.

Entrances to rooms or areas where there are live parts must have a sign posted, forbidding unqualified persons from entering.

All disconnecting means (service, feeder, or branch circuit) must be marked, showing its purpose. (This is not required if the purpose of the disconnecting means is obvious.)

Exposed parts of high- and medium-voltage systems must have adequate clearance above working spaces. *Table 110-34(e)* lists this distances.

Once again, make sure that you not only understand these requirements but also are able to find them by yourself. I have purposely left off section numbers, so that you will learn how to find things by yourself. Remember, although this is a small section, it lays the foundation for the rest of the Code. *Make sure you understand it thoroughly.*

Chapter Questions

1. How much free space is required around electrical panels?

2. What is the most basic installation requirement?

3. What must you do to an unused opening in a box?

4. What must be done to soldered splices before they are soldered?

5. What does the term *listed* mean?

6. Over what voltage must live parts be guarded?

7. Is brazing an acceptable method of splicing?

8. Is installing equipment in a room with restricted access acceptable as a method of guarding?

9. What type of insulation must a splice be covered with?

CHAPTER
4

Branch Circuit and Feeder Calculations

Article 220 of the *NEC*® gives calculations for branch-circuit and feeder loads, and methods for determining the required number of branch circuits. This article lays out the minimum requirements for electrical designs and applications. In almost all cases, however, you will want to design electrical systems that exceed these minimum standards, to make allowance for additional future uses.

VOLTAGES

The following voltages are the only ones used in calculations involving branch and feeders. This will hold true unless other voltages are specified. The voltages used are 120, 120/240, 208/120Y, 240, 480/277Y, 480, and 600 volts, nominal.

COMPUTATION OF BRANCH CIRCUITS

The loads of branch circuits must be calculated as follows:

Continuous Loads. A branch circuit's rating must be no less than 100 percent of the intermittent load, and 125 percent of the continuous load. This is excepted for continuous devices listed for continuous operation, which may be calculated at a 100 percent rating.

Lighting Load for Listed Occupancies. The calculations for loads in various occupancies are based on watts (volt-amperes) per square

foot. This is a minimum basis, and consideration should be given to the ever increasing trend toward higher levels of illumination. Each installation should be examined and not figured entirely on the watts-per-square-foot basis, but on the anticipated figure demands upon the system. The watts-per-square-foot figures are listed in *Table 220-3(b)* of the Code.

In figuring the watts per square foot (0.093 sq m), the outside dimensions of the buildings are to be used. They do not include the open porches and attached garages with dwelling occupancies. If there is an unused basement, it should be assumed that it will be finished later and should, therefore, be included in the calculations so that the capacity of the wiring system will be adequate for future service.

The unit values given are based on a factor of 100 percent for the minimum requirements, so any low-power factors should be taken into account in the calculations. If high-power-factor discharge lighting is not used, allowance should be made for the increased amperage due to the lower-power factor.

In calculations for dwellings, when figuring the area to be covered, porches, garages, and (if not adaptable for future use) unused or unfinished spaces are not to be included.

Other Loads—All Occupancies. For any lighting other than general illumination, and for appliances other than motors, minimum unit loads per outlet are given in *Section 220-3(c)*.

In this section of the *NEC®* you will find five categories necessary in making calculations. Receptacle outlets are not to be considered as less than 180 volt-amperes for each receptacle. If more than one receptacle is contained in an outlet, each receptacle must be independently rated at 180 volt-amperes. For instance, an outlet with two receptacles in it would be rated at 160 volt-amperes (80 volt-amperes for each of the two receptacles).

Exception No. 1: Multiple outlet assemblies are available, and if the multiple outlets are 5 feet or less from each other, then the continuous length of the multiple outlet assembly shall be considered as one outlet and figured at 180 volt-amperes capacity. Where a number of appliances will be used simultaneously, each foot (305 mm) will be considered as an outlet of not less than 180 volt-amperes capacity. These requirements do not apply to dwellings or to guest rooms in hotels. An example of a location where 1 foot might be considered as 180 volt-amperes capacity is an appliance sales floor where a number of appliances might be connected for demonstration.

Exception No. 2: Household electric ranges are not subject to this regulation, but their minimum loads may be determined by using *Table 220-19*.

Exception No. 3: For show-window lighting, a minimum of 200 volt-amperes per linear foot (305 mm) of show window, measured horizontally along the base of the window, shall be used.

Exception No. 4: The loads of outlets serving telephone exchanges do not have to be included in the calculations.

Exception No. 5: Electric clothes dryer loads can be calculated according to *Section 220-18*.

Loads for Additions to Existing Installations. Additional installations to existing electrical systems shall conform to the following:

1. **Dwelling Units.** When computing additional loads, for structural additions onto an existing dwelling unit, the methods mentioned above shall be used in the computation of the additional new load. Loads for structural additions to an existing dwelling unit or to a previously wired portion of an existing dwelling that exceeds 500 square feet (46.5 sq m) shall be in as required for lighting loads for listed occupancies.

2. **Other than Dwelling Units.** When adding new circuits or extensions of circuits in other than dwelling units, either volt-amperes per square foot or amperes per outlet may be used as covered.

BRANCH CIRCUITS REQUIRED

Section 220-3 of the Code covers branch circuits for lighting and for appliances, including motor-operated appliances. There will be branch circuits required elsewhere in the Code that are not covered in *Section 220-3*. Small appliance loads are covered in (b) below; laundry loads, in (c).

(a) Number of Branch Circuits

The total computed load and the size or rating of the branch circuits used will determine the number of branch circuits required. The maximum load on any branch circuit should not exceed the maximum specified in *Section 210-22*, and it is good practice to leave some spare capacity, as in most cases additional loads will be needed at some time.

(b) Small Appliance Branch Circuits — Dwelling Unit

Besides the branch circuits required by *Section 220-4(a)*, two or more 20-amp small appliance circuits are required, to accommodate the connection of portable small appliance loads.

(c) Laundry Branch Circuits — Dwelling Unit

In addition to the number of branch circuits required, including the small appliance circuits, there shall be a minimum of at least one 20-ampere branch circuit to supply laundry receptacle outlet(s) required by *Section 210-52*. There can be no other load on this circuit.

(d) Load Evenly Proportioned Among Branch Circuits

When computing on a volt-amperes-per-square-foot (0.093-sq m) basis, the load should be distributed as evenly as possible between the required circuits, and according to their capacities.

Examples of computations are found in *Chapter 9* of the *NEC*®.

FEEDERS

Ampacity and Computed Loads

In no case can the ampacity of the feeders be smaller than required to serve the load, and also in no case can it be smaller than the sum of the loads of the branch circuits that it supplies, as computed in Part A of *Article 220*. This statement is subject to applicable demand factors that may apply in Parts B, C, and D of *Article 220*.

See examples in *Chapter 9* of the Code, and see *Section 210-22(b)* for maximum load permitted, at 100 percent power factor, for lighting fixtures.

Continuous and Noncontinuous Loads

Where a feeder supplies continuous loads or any combination of continuous and noncontinuous loads, neither the rating of the overcurrent device nor the ampacity of the feeder conductors shall be lower than the ampacity of the noncontinuous load plus 125 percent of any continuous load on the feeders.

Exception: If the overcurrent devices that protect feeders are listed at 100 percent operation of their rating, the ampacity of the overcurrent protection and feeder shall be not less than the ampacity of the noncontinuous load plus the sum of the continuous load.

GENERAL LIGHTING

The demand factors listed in *Table 220-11* of the Code cover that portion of the total branch-circuit loads that is computed for lighting loads. These demand factors are only for the purpose of determining feeders to supply lighting loads, and are not for the purpose of figuring the number of branch circuits. The number of branch circuits has been covered in the preceding part, and will also be covered in *Chapter 9* of the Code, as well as in parts of the following coverage of *Article 220*.

Each installation should be specifically analyzed, since the demand factors listed in this section of the Code are based only on the minimum requirements of load conditions and for 100 percent power-factor conditions. There will be conditions of less than unity power factor and conditions where these demand factors would be wrong. With the trend to higher illumination intensities and the increased use of fixed and portable appliances, the loads imposed on the system are very likely to be greater than the minimums. Also, electric discharge lighting should be of the high-power-factor type; if not, additional provisions should be made for the low-power factor involved.

Demand factors for small appliances for laundry equipment in dwellings is covered in *Section 220-16*.

SHOW-WINDOW LIGHTING

Show-window lighting is a separate load from that calculated on the volt-amperes-per-square-foot basis. This lighting is to be figured at a minimum of 200 volt-amperes per linear foot (305 mm), measured horizontally along the base.

Branch circuits supplying show windows are covered in *Exception No. 3* of *Section 220-3(c)*.

RECEPTACLE LOADS— NONDWELLING UNITS

For installations other than dwelling units, receptacle loads calculated at 180 volt-amperes or less are allowed to be added to lighting loads, and are derated in accordance with *Tables 220-11* or *220-13*. You may also refer to *Table 220-13* for demand factors for nondwelling receptacle loads.

MOTORS

The computing of motor loads is covered in *Sections 430-24, 430-25, and 430-26.*

FIXED ELECTRIC SPACE HEATING

With two exceptions, the computed load on a feeder that serves fixed electrical space heating must be equal to the total heating load on all of the branch circuits. There is no demand factor. The feeder load current rating shall never be computed at less than the rating of the largest branch-circuit supplied.

Exception No. 1: The inspection authority enforcing the Code may grant special permission to issue a demand factor for electrical space heating where space heating units have duty-cycling or where all units will not be operating at the same time. The feeders are required to be of sufficient current-carrying capacity to carry the load as so determined.

Exception No. 2: There are optional calculations for single-family dwellings, etc. in *Sections 220-30* and *220-31*. These optional calculations shall be permitted for fixed space heating in single-family dwellings or in individual apartments of multifamily dwellings. Where the calculations are for a multifamily dwelling, *Section 220-32* will be permitted to be used.

SMALL APPLIANCE AND LAUNDRY LOADS—DWELLING UNIT

Small Appliance Circuit Load

The requirements for small-appliance receptacle outlets in single-family dwellings, multifamily dwellings with individual apartments

having cooking facilities, and in hotels or motels having serving-pantry facilities or other cooking facilities, shall be a minimum of not less than two 2-wire small appliance circuits, as required in *Section 220-4(b)*. The calculated load for each circuit shall not be less than 1500 volt-amperes. These loads may be included with the general lighting load, which makes it subject to the demand factors as shown in *Table 220-11*.

Laundry Circuit Load

A minimum of one 20-ampere laundry circuit is required in the laundry room, but a feeder load of not less than 1500 volt-amperes shall be included for each 2-wire laundry branch circuit installed as required in *Section 220-4(c)*. These circuits may be included with the general lighting load and are subject to the demand factors provided in *Section 220-11*.

APPLIANCE LOAD— DWELLING UNITS

A demand factor of 75 percent of the nameplate rating is permitted where there is a load of four or more fixed appliances served by the same feeder in a single-family dwelling or a multifamily dwelling.

Exception: The demand factors above do not apply to electric ranges, clothes dryers, air-conditioning, or space heating equipment.

ELECTRIC CLOTHES DRYERS— DWELLING UNITS

Electric clothes dryer units shall be calculated at 5000 watts (volt-amperes) or the nameplate rating on the dryer, whichever is larger. When a house is being wired, the wireman often cannot be certain what the rating of the clothes dryer to be purchased will be. This is why Code has placed the 5000-watt minimum. Use of the demand factors in *Table 220-18* are permitted.

Many inspectors have insisted on No. 8 copper or the equivalent for dryers, as most dryers draw 30 amperes or more. Nameplate rating and a branch circuit should not be loaded to exceed 80 percent of its capacity.

ELECTRIC RANGES AND OTHER COOKING APPLIANCES — DWELLING UNITS

In calculating feeder loads for electric ranges or other cooking appliances in dwelling occupancies, any that are rated over 1-3/4 kW shall be calculated according to *Table 220-19* in the *NEC*®. The notes following *Table 220-19* are a part of the table and are very important in the calculation of feeder and branch-circuit loads.

Due to the increased wattages being used in modern electric ranges, it is recommended that the maximum demands for any range of less than 8-3/4 kW rating be figured using Column A in *Table 220-19*.

Three-phase, 4-wire wye systems are often used. When calculating the current in such systems, it is necessary to use a demand that is twice the maximum number of ranges that will be connected between any 2-phase wires. An example of this is shown under Example 6 in *Chapter 9* of the *NEC*®.

KITCHEN EQUIPMENT — OTHER THAN DWELLING UNITS

Table 220-20 of the *NEC*® is used to compute the load for commercial electric cooking equipment and associated equipment (dishwashers, water heaters, etc.). The demand factors shown in this table apply to all equipment that is either controlled by thermostats or intermittent in operation. These demand factors do not apply to electric heating or air-conditioning systems. Note that the demand can never be less than the combined load of the two largest pieces of equipment.

NONCOINCIDENTAL LOADS

When adding branch circuit loads, the smaller of any two loads can be omitted if it is almost certain that both will not be used at the same time.

FEEDER NEUTRAL LOAD

In general, the feeder neutral load is determined by the maximum unbalanced load. This must be considered whenever a neutral is used by two or more ungrounded conductors.

For feeders that supply electric ranges, the maximum neutral load is considered to be 70 percent of the load on the ungrounded conductors. Therefore, the neutral conductor in a feeder that supplies an electric range must be sized at least 70 percent as large as the ungrounded conductors, but no smaller than No. 10 copper. (Wire sizes in the Code always refer to copper conductors, unless specifically stated otherwise.)

NEUTRAL SIZING FOR ELECTRIC DISCHARGE LIGHTING

Because of the electrical characteristics of electric discharge lighting, the neutrals that feed power to these circuits must be sized as large as the ungrounded conductor(s). Due to what is called a *third harmonic* (a 20-cycle current caused by electrical and magnetic characteristics of these lighting fixtures and their ballasts), the neutrals that supply these circuits sometimes carry as much current as the ungrounded (hot) conductors.

SAMPLE CALCULATION

Here are the calculations for a 50-unit apartment complex in which each apartment is about 1000 square feet. The power is supplied with a 3-phase, 4-wire, 120/208-volt wye system:

General lighting,
  50 apts × 3 watts × 1000 sq ft = 150,000 VA
Appliance loads,
  50 apts × 2 ckts × 1500 watts = <u>150,000 VA</u>
                                        300,000 VA

3,000 VA @ 100%	3,000 VA
117,000 VA @ 35%	40,950 VA
150,000 VA @ 35%	45,000 VA
Minimum feeder capacity	88,950 VA

With 50 electric ranges on a 3-phase system, there will be 16 ranges on one phase (single-phase ranges), and 17 ranges on each of the other two phases. From *Table 220-19,* we find that the maximum number of ranges connected at any one time is 34. Also according to this table, the load will be 15,000 watts + 34,000 watts, equaling 49,000 watts.

Therefore, the maximum feeder demand will be:

49,000 VA	from *Table 220-19*
88,950 VA	from previous calculations
137,950 VA	Total

The ampacity will be as follows:

$$\frac{137,950 \text{ VA}}{1.732 \times 208 \text{ V}} = 384 \text{ amps}$$

The neutral will be sized as follows:

General lighting and appliances at 100%	88,950
Electric ranges, 49,000 @ 70%	34,300
Total	123,250 VA

Total neutral current is:

$$\frac{123,250 \text{ VA}}{1.73 \times 208 \text{ V}} = 342 \text{ amps}$$

First 200 amps @ 100%	200.0 amps
Remaining 142 amps @ 70%	99.4 amps
Total	299.4 amps

Once the phase load (384 amps) and the neutral load (299 amps) have been calculated, you can use *Table 310-16* to find the correct conductor sizes for these conductors.

OPTIONS

There are a number of optional calculations given in the *NEC®.* Refer to the Code if you need to do calculations for an unusual installation. Normally these calculations are performed by engineers, but it is important that you have a basic understanding of how they are performed.

Chapter Questions

1. What factors decide how much current a conductor is allowed to carry?

2. How long must a load be continuously on before it is considered a continuous load?

3. Should attached garages be included in square-foot measurements for calculating loads?

4. What is the minimum load to be calculated for each receptacle outlet?

5. How many volt-amps (VA) must be calculated for each linear foot of show windows?

6. How many small appliance circuits are required in a new home?

7. What should the calculated load be for the circuits mentioned in number 6?

8. Circuit breakers that handle continuous loads should be rated at what percentage of the load?

9. What is the feeder demand factor for four or more fixed appliances in a dwelling?

10. What is the minimum load that can be calculated for an electric clothes dryer?

CHAPTER
5

Overcurrent Protection

Overcurrent protection is the part of an electrical system that is most important in preventing electrical fires. Fires are caused by excessive levels of heat, and the amount of heat produced by electrical currents is dependent upon how much current is flowing. For example, if a No. 14 wire carries 10 amps of current, it may be warm, but it is certainly not dangerously hot. If, however, you attempt to run 40 amperes of current through it, it will become dangerously hot, and in many instances could cause a fire. In scientific terms, we can say that the amount of heat produced by a conductor is *directly proportional* to the amount of current flowing through the conductor—more current, more heat; less current, less heat.

Because of this fact, we need to limit the amount of current flowing through conductors so that we can be sure that the conductors will not get hot enough to become a hazard. We call our methods of doing so *overcurrent protection,* protection against overly high currents.

CONDUCTOR SIZES

When applying overcurrent protection, we must carefully consider the size of the conductors that will be carrying the currents. This is important because some conductors can carry more current safely than others. The determining factors are size and material. All electrical conductors heat up when currents flow through them, with size and material differences determining how much.

The larger a conductor is, the more current it can carry without overheating. This size is measured in cross-sectional area (the amount of area the conductor covers at a cross section), usually stated in AWG gauge or in thousand circular mils (kcmil).

The material of the conductor is also important. In general, the only materials used for electrical conductors are copper and aluminum. Of these two, copper is the better conductor. When aluminum conductors are used, they must be about 40 percent larger than a copper conductor to carry an equal amount of current.

Since insulation is used on most electrical conductors, the type of insulation used becomes another factor. The conductor must not heat up any further than the insulation can stand.

Also critical is the surrounding area. A conductor outdoors in winter can carry a lot more current than one installed in a conduit indoors. The critical factor here is the conductor's ability (in the particular circumstance mentioned) to dissipate heat. This depends on the temperature of the conductor's surroundings (called *ambient temperature*) and the amount of airflow around the conductor, which removes heat from the area.

You will find that the *NEC®* takes all of these factors into consideration, and that some skill is required to properly apply overcurrent protection.

Overcurrent protection is provided by the careful use of fuses and circuit breakers, which automatically cut off current to a circuit if the current level rises too high. These devices are extremely reliable, but must be installed properly. If they are incorrectly used, they are of very little value, and are perhaps more dangerous than no protection, since they may give an illusion of safety.

The requirements of *Article 240*, which covers overcurrent protection, are as follows:

All conductors (except for some flexible cords) must have overcurrent protection no greater than their ampacity.

There are modifications to the above requirements for specific types of circuits (such as some motor circuits, some remote-control circuits, etc.), which are covered by their specific articles.

The standard ampere ratings for fuses and inverse time circuit breakers are as follows: 15, 20, 25, 30, 35, 40, 50, 60, 70, 80, 90, 100, 110, 125, 150, 175, 200, 225, 250, 300, 350, 400, 450, 500, 600, 700, 800, 1000, 1200, 1600, 2000, 2500, 3000, 4000, 5000, and 6000. Fuses are also rated 1, 3, 6, 10, and 601 amps.

For the purpose of making calculations, adjustable trip circuit breakers are considered as rated at their maximum possible setting.

Fuses or circuit breakers cannot be connected in parallel unless factory assembled in parallel.

Supplementary protection, such as in-line fuses installed in lighting fixtures, cannot replace branch-circuit overcurrent protection.

No overcurrent device can be connection in series with any grounded conductor, unless the overcurrent device opens all conductors of the device, so that no conductor can operate independently.

When a change in size is made in the ungrounded (hot) conductor, a similar change can be made in the grounded (neutral) conductor.

Except where specifically excepted, overcurrent devices must be accessible.

Overcurrent devices must be connected at the point where the circuit to be protected receives its supply. (See Figures 5-1 and 5-2.)

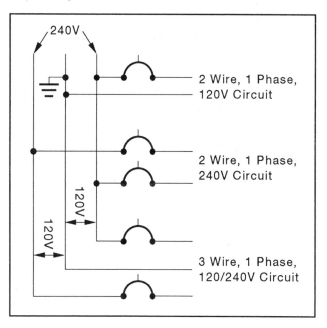

Fig. 5-1 Overcurrent devices in circuits from single-phase system.

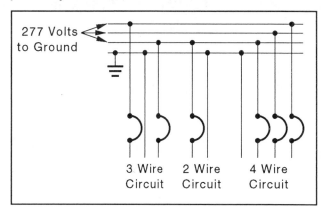

Fig. 5-2 Overcurrent devices in circuits from 3-phase system.

TAPS

The requirement that all overcurrent devices must be connected at the point where the circuit receives its supply is excepted for taps. Taps are used primarily in industrial installations, where conditions require their use. They are almost never used in residential applications, and only sometimes in commercial ones. It is recommended that you avoid the use of taps unless you encounter circumstances that require them. In such cases, the following requirements apply:

If tapes are to be made from feeder (supply) conductors, they must:

1. Be no longer than 10 feet. (See Figure 5–3.)
2. Terminate in an overcurrent protective device.
3. Have an ampacity no less than the device being fed or the overcurrent device mentioned above.
4. Extend no further than the device they supply.
5. Be enclosed in a raceway.
6. Have an amp at least 10 percent of the fuse or circuit breaker on the line side.

Feeder taps up to 25 feet long can be made under the following conditions (see Figure 5–4):

1. The tap must terminate into a branch-circuit protective device.
2. The tap must have an ampacity of at least one-third of the feeder ampacity. (For taps made to transformer secondaries, the secondary must have an ampacity that, when multiplied by the secondary-to-primary voltage ratio, is at least one-third of the ampacity of the conductors or overcurrent device from which the primary conductors are tapped.)
3. The tap must be protected from physical damage.

In high-bay manufacturing buildings (more than 35 feet from floor to ceiling, measured at the walls), taps longer than 25 feet are permitted. In these cases, the tap conductors must:

1. Have an ampacity of at least one-third that of the feeder conductors.
2. Terminate in an appropriate circuit breaker or set of fuses.
3. Be protected from damage and installed in a raceway.
4. Be continuous with no splices.
5. Be a minimum size of No. 6 AWG copper or No. 4 AWG aluminum.

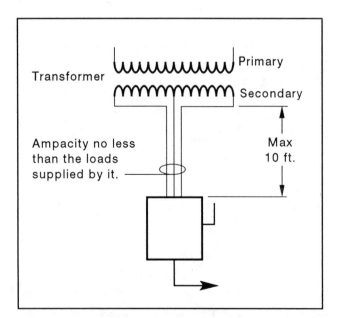

Fig. 5–3 Basic tap circuit.

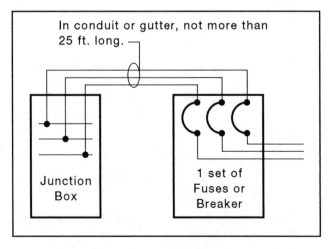

Fig. 5–4 Transformer tap.

6. Not penetrate floors, walls, or ceilings.
7. Run no more than 25 feet horizontally or 100 feet overall.

Tap conductors are allowed from a 50-ampere circuit to a range, wall-mounted oven, cook-top, or other household cooking appliance. They must run no longer than necessary, and must be rated at least 20 amps.

Circuits with no grounded conductors are allowed to be tapped from circuits with grounded conductors. Switching devices in the tapped circuits must have a pole in each ungrounded conductor. If multipole switches function as a disconnecting means, all conductors must open simultaneously when the device is activated.

Tap conductors are allowed from a 40- or 50-ampere circuit to loads other than household cooking appliances. They must run no longer than necessary, and must be rated at least 20 amperes.

Tap conductors are allowed from a circuit under 40 amperes to loads other than household cooking appliances. They must run no longer than necessary, and must be rated at least 15 amperes.

All branch-circuit or feeder taps from busways must use plug-ins or connectors that have overcurrent devices in them.

The above requirements apply in the following circumstances:

1. Where fixtures that have no overcurrent device mounted on them are mounted directly onto the busway.
2. Where an overcurrent device is part of a cord plug for cord-connected fixed or semifixed lighting fixtures.

For group motor installations, taps to single motors do not need branch-circuit protection in any of the following cases:

1. The conductors to the motor have an ampacity equal to or greater than the branch-circuit conductors.
2. The conductors to the motor are no longer than 25 feet, they are protected, and the conductors have an ampacity at least one-third as great as the branch-circuit conductors.

Secondary conductors can be tapped from transformers of separately derived systems when the following conditions are met:

1. The conductors are not longer than 25 feet. (See Figure 5-5.)
2. All overcurrent devices are grouped.
3. The tap conductors are protected.

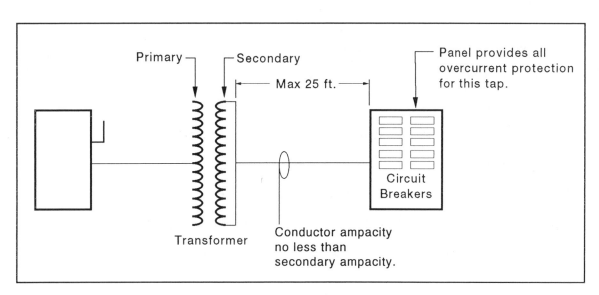

Fig. 5-5 25-foot transformer tap.

4. The ampacity of the conductors is no less than the secondary current rating of the transformer and the sum of the overcurrent devices will limit this capacity.

LOCATIONS

Obviously, the areas in which overcurrent protective devices are installed will have a marked effect upon their operation. This is especially important because many of them operate thermally. That is, their operation is triggered by heat.

The requirements for installation locations for overcurrent devices are as follows:

Overcurrent devices must be located where they will not be subjected to physical damage.

Overcurrent devices should be enclosed in cabinets or enclosures and mounted in the vertical position (unless impractical).

When installed in damp or wet locations, overcurrent devices must be mounted with at least a 1/4-inch air space between the enclosure and the surface on which they are mounted.

MISCELLANEOUS

Edison-base fuses are used for replacement only. When new fuses are installed, they should be Type S. This requires the installation of the Type S shell (which simply screws into the fuse socket), so only Type S fuses can be used. Since Type S fuses and shells come in 15-, 20-, and 30-amp configurations and cannot be used interchangeably, this guarantees that only the appropriate size of fuse can be used, once the S shell is in place. This system was developed to eliminate the hazards associated with replacing a blown fuse with a fuse of a higher value. There have been many fires caused by persons replacing 15-amp fuses (that were protecting 14-gauge wire) with 30-amp fuses.

A disconnecting means must be installed on the line side of fuses operating at more than 150 volts to ground. (Disconnect switches are built this way.)

Circuit breakers to be used as switches must be marked "SWD."

Chapter Questions

1. What causes conductors to become hot?

2. What are the two most common types of overcurrent protective devices?

3. Why must aluminum conductors be sized larger than copper conductors?

4. What is ambient temperature?

5. What is ampacity?

6. Is 55 amps a standard ampere rating for circuit breakers?

7. Can in-line fuses be used instead of circuit breakers?

8. Must overcurrent protective devices be accessible?

9. How much of an air gap is required for overcurrent devices in wet locations?

10. What type of plug fuses must be used for new installations?

CHAPTER
6

Wiring Requirements

Article 300 of the *NEC*®sets forth the basic requirements for wiring. The requirements of this section of the *NEC*® are for wiring systems of 600 volts or less, and might not apply to the internal conductors of motors, controllers, machinery, etc. Basically, it applies to all types of wiring not covered by other parts of the *NEC*®. Almost all types of wiring must comply with this article. Even wiring methods specified in other parts of the *NEC*® usually have to comply with many of the basic requirements of this article.

These are the basic requirements that specify exactly what is a safe electrical wiring installation. This is one of the more important sections of the Code, although it is not used as often as some others.

GENERAL REQUIREMENTS

Single conductors are only permitted where they are part of a standard wiring method.

All ungrounded conductors, neutral conductors, and equipment grounding conductors of one circuit must be run in the same raceway, cable, cable tray, trench, or cord.

AC or DC circuits of voltages not over 600 volts are allowed in the same raceway, cable, or enclosure. However, all wires in the raceway, cable, or enclosure must have sufficient insulation for the highest voltage present in the raceway, cable, or enclosure. (120-volt circuits can occupy the same raceway with 600-volt circuits [AC or DC], as long as all conductors in the raceway are insulated for 600 volts.)

Circuits over 600 volts and circuits 600 volts or less cannot occupy the same raceway, cable, or enclosure. (Exceptions are made for lighting fixture ballast wires and control instrument wires.)

PROTECTING CONDUCTORS

For obvious reasons, it is very important that conductors (and especially their insulation) are carefully protected. The following requirements are designed to make sure that all conductors are safely protected:

All conductors must be protected from physical damage.

When cables or raceways are installed through wood structural members, the edge of the bored holes must be at least 1-1/4 inches from the edge of the wood framing member.

Where the clearance specified above is not possible, a 1/16-inch steel plate must be installed to cover the area of the wiring. (See Figure 6-1.)

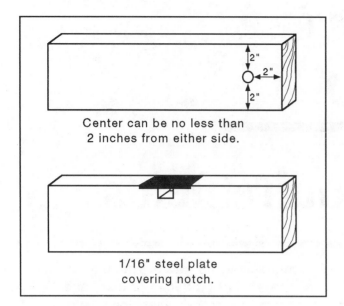

Fig. 6-1 Requirements for notching or drilling framing members.

Notches can be cut in wood if the strength of the structure will not be jeopardized. Wiring can be laid in the notches, and a 1/16-inch steel plate must be installed to cover the area.

The steel plate mentioned above for covering notches is not required where the wiring method is one of the following:

1. Rigid metal conduit.
2. Intermediate metal conduit.
3. Electrical metallic tubing.
4. Rigid nonmetallic conduit.

Where nonmetallic sheathed cables go through metal framing members (usually metal studs), a grommet or bushing must protect the hole the cable passes through.

In locations where nails or screws are likely to damage nonmetallic sheathed cables or electrical nonmetallic tubing, a 1/16-inch steel plate must be installed to cover the area.

When cables or raceways are installed parallel with structural members, the edge of the cable or raceway must be at least 1-1/4 inches from the edge of the framing member.

Where the clearance specified above is not possible, a 1/16-inch steel plate must be installed to cover the area of the wiring.

The steel plate mentioned above is not required where the wiring method is one of the following:

1. Rigid metal conduit.
2. Intermediate metal conduit.
3. Electrical metallic tubing.
4. Rigid nonmetallic conduit.

UNDERGROUND WIRING

Buried conductors present different hazards from those installed above ground. Chiefly these hazards are those of damage from digging tools, constant wetness (in many locations), mechanical stresses where the conductors leave the ground (going up into a building, conduit, etc.), and animals who chew on the cables.

The following requirements were written to address all of the following possible hazards:

Directly buried cables or conduits must meet the depth requirements shown in *Table 300-5*. (See Figures 6-2 through 6-5.)

Underground cables installed beneath a building must be in raceway the full length of the building.

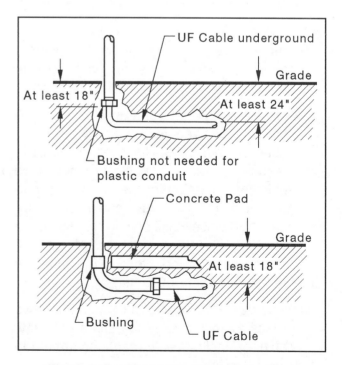

Fig. 6-2 Depth requirements for cables.

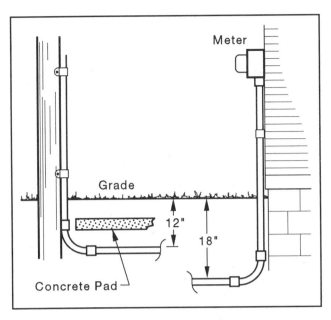

Fig. 6-3 Depth requirements for conduits.

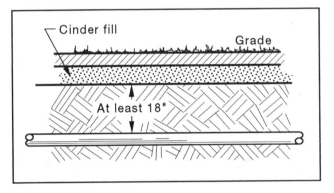

Fig. 6-4 Conduit run under cinder fill.

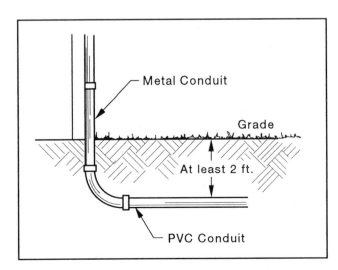

Fig. 6-5 Change from metal conduit to PVC conduit.

Conductors or cables directly buried must be protected from below where they emerge from the ground to a point 8 feet above the grade level. The protection can be provided by an enclosure or raceway.

Conductors that enter a building must be protected up to the point of entrance.

Directly buried conductors are allowed to be spliced without a junction box. However, the splicing method must be approved for the intended use.

Care must be taken so that backfill will not damage conductors. When necessary, a layer of sand or gravel should be placed over the conductors to protect them.

Conduits or raceways that could allow moisture to enter a building must be sealed.

Cables that terminate underground (with directly buried wiring protruding from them) must be terminated with a bushing or an equivalent sealing compound.

All conductors of the same circuit must be installed in the same raceway or, in the case of open conductors or cables, close to each other in the trench. Cables in trenches must be close to each other for two main reasons: First, because they are less likely to be damaged if they are grouped together, and secondly, because of the effects of mutual inductance. Mutual inductance, an advanced concept not covered in this text, basically means that not grouping conductors together can increase the resistance of the circuit.

Parallel conductors are allowed, but each raceway must contain all the conductors of the same circuit.

All metal raceways, cables, and fittings must be suitable for the area in which they are installed. If corrosive conditions exist, raceways must be covered with a corrosion-resistant coating or other suitable protection.

In wet locations, and in locations where the walls are frequently washed (dairies, laundries, etc.), all raceways, cables, enclosures, etc. must be installed with an air space of at least 1/4 inch between them and the surface they are mounted upon.

The air space requirement mentioned above also applies when raceways, cables, or enclosures are to be mounted on absorbent materials such as damp paper or wood.

RACEWAYS

Remember that when the Code refers to raceways, it is not referring to conduits only. While conduits are by far the most common type of raceway, there are a number of other types as well. The definition of the term *raceway* is found in *Article 100* of the *NEC®*, which states that a raceway is "An enclosed channel designed expressly for holding wires, cables, or busbars, with additional functions as permitted in this Code."

The requirements for raceways are as follows:

Airflow through a raceway must be prevented (usually by sealing) if the two ends of a run are exposed to noticeably different temperatures. (This applies frequently in commercial freezer installations.)

Raceways or cable trays that contain electric wiring cannot also contain any other type of piping (steam, gas, air, drainage, etc.).

All raceways, cables, and boxes must be electrically and mechanically joined to-

gether, except as allowed for nonmetallic boxes. (See Figure 6-6.)

All raceways, cables, and boxes must be securely supported.

Raceways are not allowed to support other raceways, cables, or nonelectric equipment, unless identified as suitable for such use. (Class 2 cables can also be supported if they are used only for equipment control circuits.)

No splices are allowed in raceways.

The continuity of a grounded conductor cannot be dependent upon device connections. (Neutrals cannot "feed through" a wiring device but must be made up with "pigtail" connections.)

At least 6 inches of free conductor must be left at every box, fitting, or other splice point.

The above requirement does not apply to conductors that do not splice or terminate in the box but rather feed through.

BOXES

While boxes are covered in detail in *Article 370,* the foundational requirements for boxes are given in *Article 300.* Specifically, the following requirements are given in *Sections 300-15* and *300-16:*

Boxes or fittings are required at every splice or pull point in raceways (wireways, gutters, and cable trays excepted).

Boxes or fittings are required at every splice or pull point for Type AC, MC, MI, NM, or other cables. (Certain exceptions exist. If in doubt, refer to the exceptions in *Section 300-15(b).)*

Fittings, etc. can only be used with the system they were designed for. (No Greenfield fittings on conduit, etc.)

A box or fitting with a separately bushed hole for each conductor must be used

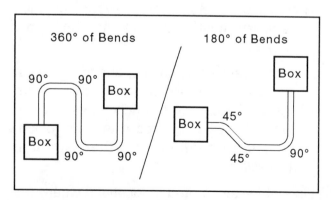

Fig. 6-6 Degrees of bend in conduit runs.

where a change is made between raceway or cable systems and open or knob-and-tube wiring methods.

Where a raceway terminates behind a switchboard, a bushing on the end of the raceway can be used instead of the box with bushed holes mentioned above.

GENERAL REQUIREMENTS FOR CONDUCTORS

The conductor requirements of *Article 300* are not the only conductor requirements in the *NEC*®. All of *Article 310* is devoted to conductors, as well as a number of other sections. These requirements are the basic safety requirements. In many other sections of the Code, additional requirements are given for specific circumstances.

The general requirements are as follows:

The number and size of conductors in raceways must not be too great to allow for heat dissipation and ease of installation and removal.

Raceways must be completely installed before conductors are pulled into the raceway system.

Conductors installed in vertical raceways must be supported. The support requirements are shown in *Table 300-19(a)*.

Conductors of AC systems must be grouped together in circuits to avoid induction heating.

Where single AC conductors are used, the inductive effects must be avoided by the following methods:

1. Cutting notches in the metal between the holes the conductors pass through.
2. Running the conductors through a box or enclosure wall made of insulating material.

Openings around electrical penetrations in fire-rated walls, ceilings, floors, etc. must be properly sealed against the spread of fire.

No wiring of any type is allowed in ducts that transport dust, loose stock, or flammable vapors.

The following wiring methods can be used in ducts or plenums used for environmental air:

1. Rigid metal conduit.
2. Intermediate metal conduit.
3. Electrical metallic tubing.
4. Flexible metallic tubing.
5. Type MI cable.
6. Type MC cable that has a nonmetallic covering.
7. Flexible metal conduit, no longer than 4 feet.
8. Flexible liquidtight metal conduit, no longer than 4 feet.

Equipment and devices are allowed in environmental air and plenum spaces only when necessary.

The following wiring methods can be used in spaces other than ducts or plenums used for environmental air:

1. Rigid metal conduit.
2. Intermediate metal conduit.
3. Electrical metallic tubing.
4. Flexible metallic tubing.
5. Type MI cable.
6. Type MC cable that has a nonmetallic covering.
7. Flexible metal conduit.
8. Flexible liquidtight metal conduit, no longer than 6 feet.
9. Type AC cable, or other listed cables.
10. Totally enclosed, nonventilated busway, with no provision for plug-ins.

Other types of conductors and cables can be installed in spaces other than ducts or plenums used for environmental air in the following:

1. Cable tray systems; solid bottom with solid metal covers; only where accessible.

2. Surface metal raceways or metal raceways with removable covers; only where accessible.

Electrical equipment with metal enclosures or electrical equipment in listed nonmetallic enclosures can be installed in the places mentioned above. Integral fan systems can also be installed in such places.

Wiring under raised floors in data processing areas must meet the requirements for such areas. (See *Article 645.*)

Chapter Questions

1. AC and DC circuits below what voltage can be installed in common raceways?

2. How far from an edge must cables be when installed through 2 × 4s?

3. What thickness of steel plate can be used to protect wiring passing through structural members?

4. What are grommets used for with cables and metal studs?

5. In what table are the depth requirements for underground wiring given?

6. What must be prevented when the two ends of a raceway are subjected to temperature differences?

7. Are special fittings required for terminating Type AC cables in a box?

8. Why must raceways not be filled too full?

9. Why must penetrations in firewalls be sealed?

10. When are electrical devices permitted in plenum spaces?

CHAPTER
7

Grounding

Along with overcurrent protection, grounding is critical for safety. Essentially, the two-fold purpose of grounding is to provide a reliable return path for errant currents and to provide protection from lightning.

At first, it may seem that it would be better not to provide a good return path, making it harder for errant currents to flow. While this method would certainly be effective in reducing the size of *fault currents* (currents that flow where they are not intended), it would also allow them to flow more or less continually when they occur.

Since it is these fault currents that pose the greatest danger to people, our primary concern is to eliminate them entirely. We do this by providing a clear path (one with virtually zero resistance—a "dead" short circuit) back to the power source, so that these currents will be very large and thus activate the fuse or circuit breaker. The overcurrent protective device will then cut off all current to the affected circuit, eliminating any danger. This also ensures that the circuit cannot be operated while the fault is present, which makes speedy repairs unavoidable.

To put things in simple terms, we could say that we use grounding to make sure that when circuits fail, they fail all the way. Partially failed circuits are the ones that are dangerous because they can go unnoticed; therefore, grounding is essential.

HOW THE CODE COVERS GROUNDING

The main requirements for grounding are found in *Article 250,* although other sections of the code can have grounding requirements that apply only to certain specific circumstances. Remember this: Unless it is *specifically* stated otherwise, all of the requirements of *Article 250* apply to every electrical installation.

You will find in this article that there are different grounding requirements for different types of electrical systems. But while these requirements may be different for different situations, the basic idea as stated above is the desired end effect.

Many of the systems shown in this article (such as 2-wire DC systems, impedance-grounded systems, etc.) are pretty uncommon, and you may never actually work on such a system. But, on the other hand, who knows what tomorrow will bring? Besides, the Code must cover all systems that exist in commercial use.

Go through these requirements along with your Code book, and pay special attention as to how the requirements guarantee the same effects in very different systems. You will see a few instances where there are exceptions, but usually only where outside circumstances make exceptions to the rules absolutely necessary.

GROUNDING REQUIREMENTS FOR DIFFERENT TYPES OF SYSTEMS

Two-wire DC systems that supply premises wiring must be *grounded* (connected to the earth by one of the methods the Code specifies), unless one of the following situations exists:

1. The system supplies only industrial equipment in limited areas, and has a ground detector.
2. The system operates at 50 volts or less between conductors.
3. The system operates at over 300 volts between conductors.
4. The system is taken from a rectifier, and the AC supplying the rectifier is from a properly grounded system.
5. The system is a DC fire-protective signaling circuit that has a maximum current of 0.03 amperes.

The neutral conductor of all 3-wire DC systems must be grounded.

Alternating-current (AC) circuits operating at less than 50 volts must be grounded if any of the following conditions exist:

1. The circuit is installed as overhead wiring outside buildings.
2. The circuit is supplied by a transformer, and the transformer supply circuit is grounded.
3. The circuit is supplied by a transformer, and the transformer supply circuit operates at over 150 volts to ground.

AC systems operating at between 50 and 1000 volts (these are the most common systems, and the ones you are most likely to work on), and supplying premises wiring must be grounded if any of the following conditions exist (see Figure 7-1):

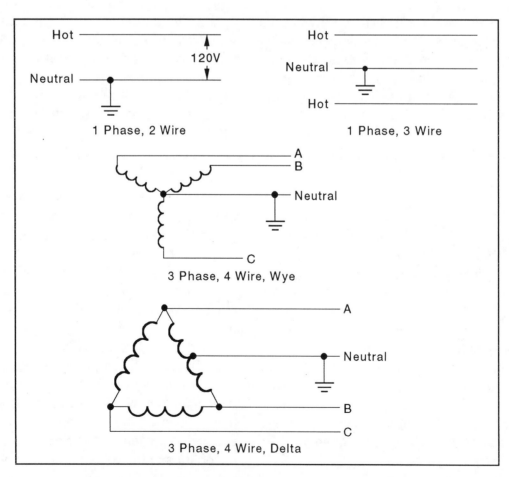

Fig. 7-1 Grounding of various types of power systems.

1. The system can be grounded so that the voltage to ground of ungrounded conductors cannot be more than 150 volts.
2. The system is a 3-phase, 4-wire wye, and the neutral is used as a circuit conductor.
3. The system is a 3-phase, 4-wire delta, and the midpoint of one phase is used as a circuit conductor.
4. When a grounded service conductor is not insulated. The following situations are excepted from this requirement:
   a. Systems used only to supply industrial electric furnaces used for melting and refining.
   b. Separately derived systems used only to supply rectifiers that supply adjustable speed drives.
   c. Other specialized systems. (See exceptions in *Section 250-5[b]*.)

Circuits for cranes operating in Class III locations over combustible fibers may *not* be grounded.

CONDUCTORS THAT MUST BE GROUNDED

The following conductors must be grounded:

1. One conductor of a single-phase, 2-wire system must be grounded.
2. The neutral conductor of a single-phase, 3-wire system must be grounded.
3. The center tap of a wye system that is common to all phases must be grounded.
4. A delta system must have one phase grounded.
5. In a delta system where one phase is used as shown above, the midtap must be grounded and must be used as the neutral conductor.

GROUNDING ELECTRODE SYSTEMS

An electrical installation's grounding electrode system connects that system to ground.

This is a critically important link in a grounding system, and requires carefully chosen materials and methods.

All of the items mentioned below are suitable for grounding electrodes. All of the following (where available) must be bonded together, forming the grounding electrode system (see Figures 7-2 through 7-5):

1. The nearest structural metal part of the structure that is grounded.

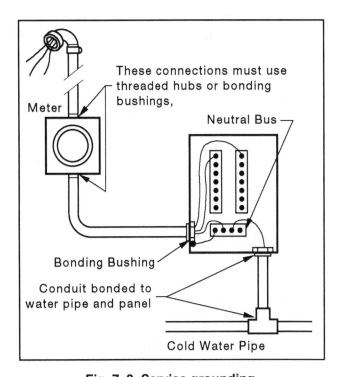

Fig. 7-2 Service grounding.

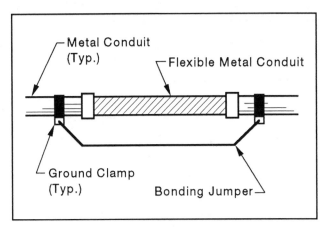

Fig. 7-3 Bonding around flexible fitting or expansion joint.

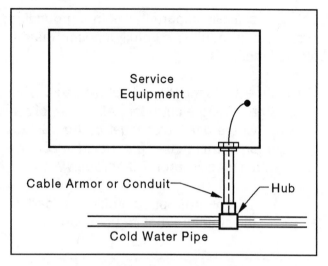
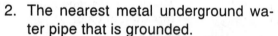

Fig. 7-4 Bonding of conduit to water pipe.

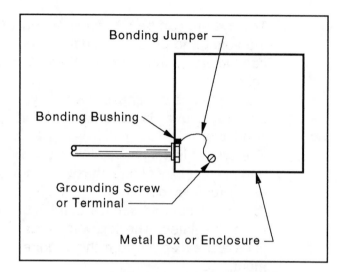

Fig. 7-5 Bonding to enclosure.

2. The nearest metal underground water pipe that is grounded.
3. An electrode (usually No. 4 minimum bare copper wire or reinforcing bar) at least 20 feet long and in direct contact with the earth for its entire length. If reinforcing bar is used, it must be at least 1/2 inch in diameter.
4. A ring of No. 2 minimum bare copper wire, at least 2-1/2 feet below grade and encircling the structure (at least 20 feet long).
5. A ground rod, plate electrode, or made electrode.

Gas piping or aluminum electrodes are not acceptable.

GROUNDING CONDUCTORS

The grounding electrode conductor is the conductor that runs between the main service disconnect enclosure and a grounding electrode (most commonly a ground rod or cold-water pipe). This is a key element in completing the ground circuit, which, as I have previously said, is critical in ensuring the safety of an electrical system.

A grounding electrode conductor can be copper, aluminum, or copper-clad aluminum. Its size must be determined by *Table*

250-94, or if the phase conductors are larger than 1100 kcmil copper or 1750 kcmil aluminum, the grounding electrode conductor must be at least 12-1/2 percent of the area of the largest phase conductor. If the service conductors are paralleled, the size of the grounding electrode conductor must be based on the total cross-sectional area of the largest set of phase conductors.

The grounding electrode conductor cannot be spliced. (Busbars and required taps are excepted.)

An equipment grounding conductor can be any of the following, and must be sized according to *Table 250-95:*

1. A copper (or other type that is corrosion resistant) conductor.
2. Rigid metal conduit.
3. Intermediate metal conduit.
4. Electrical metallic tubing.
5. Listed flexible metal conduit, in lengths of 6 feet or less.
6. The armor of Type AC cable.
7. The sheath of mineral-insulated, metal-sheathed cable.
8. The sheath of Type MC cable.
9. Cable trays.
10. Cablebus.
11. Other metal raceways that are electrically continuous.

Grounding electrode conductors can be installed in any of the following:

1. Rigid metal conduit.
2. Intermediate metal conduit.
3. Rigid nonmetallic conduit.
4. Electrical metallic tubing.
5. Cable armor.

Grounding electrode conductors that are No. 6 copper or larger can be attached directly to a building surface. (They are not required to be in a raceway unless exposed to physical damage.)

Isolated metal parts of outline lighting systems can be bonded together by a No. 14 copper or No. 12 aluminum conductor that is physically protected.

No automatic cutout or switch can be placed in the grounding conductor.

GROUNDING CONDUCTOR CONNECTIONS

The connection between a grounding electrode conductor and a grounding electrode must be accessible, permanent, and effective. (The safety of the entire system often depends on this connection.)

Where metal piping systems are used as grounding electrodes, all insulated joints or parts that can be removed for service must have jumpers installed around them. The jumpers must be the same size as the grounding electrode conductor. (See Figure 7-6.)

All grounding connections must be listed means, and must be permanent and secure.

If grounding connections are made in areas where they can be subjected to physical damage, they must be protected.

If more than one grounding conductor is present in a box or enclosure, all grounding conductors in the box must be connected. This connection must be made in such a way that the removal of any one

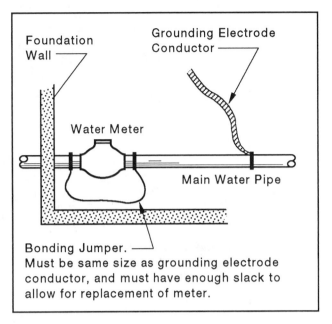

Fig. 7-6 Bonding around water meter.

device, etc. cannot affect the connection. (In common language, grounding connections cannot "feed through" but must "pigtail." See Figure 7-7.)

Metal boxes must be connected to a grounding conductor, whether it be a conduit used as the grounding conductor, or a separate grounding wire. (See Figure 7-8.)

All paint or foreign substances must be removed from the area of grounding connections. (This requirement is nothing more than common sense.)

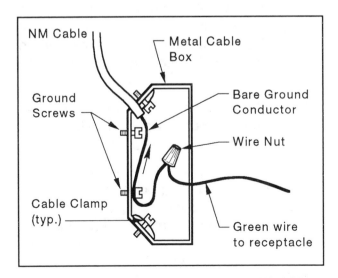

Fig. 7-7 Bonding of metal box with NM cable.

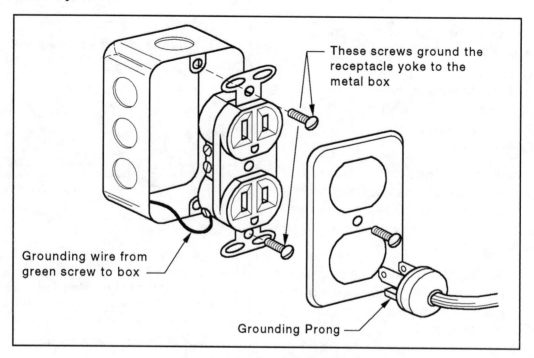

These screws ground the receptacle yoke to the metal box

Grounding wire from green screw to box

Grounding Prong

Fig. 7–8 Bonding of receptacle.

MULTIPLE GROUNDING PATHS

If the use of multiple grounding paths creates a problem (as it can in certain circumstances, such as when a number of panels have a main bonding jumper installed, sending circuit current through the grounding system), one of the following steps can be taken:

1. Discontinue one or more of the grounding locations.
2. Change the locations of the connections.
3. Interrupt the conductor or path for the objectionable ground current.
4. Use other remedies approved by the authority having jurisdiction (the local inspector).

The above rule should not be taken as allowance for not having every item connected to the system grounded.

DC systems required to be grounded must have the grounding connection made at one or more supply stations. These connections cannot be made at individual services or on premises wiring. (If the DC source is located on the premises, a grounding connection can be made at the first disconnecting means or overcurrent device.)

GROUNDING CONNECTIONS FOR AC SYSTEMS

AC systems that require grounding must have a grounding electrode conductor at each service that connects to a grounding electrode. The grounding electrode conductor must be connected to the grounded service conductor at an accessible location between the load end of the service and the grounding terminal. (See Figure 7–9.)

Where the transformer supplying the load is located outside the building, a separate connection must be made between the grounded service conductor and a grounding electrode, either at the transformer or at another location *outside* the building.

No grounding connection (to the grounding electrode) can be made on the load side of the service disconnecting means. (If this is done, it sometimes results in problematic and potentially dangerous ground-current loops.)

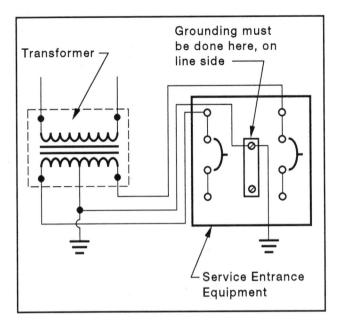

Fig. 7-9 Proper location of ground.

If an AC system that operates at less than 1000 volts is grounded, the grounded conductor must be run to every service disconnecting means and bonded to every disconnecting means enclosure. This conductor must be run with the phase conductors and cannot be smaller than the grounding electrode conductor (as specified in *Table 250-94*). If the phase conductors are larger than 1100 kcmil copper or 1750 kcmil aluminum, the grounded conductor must be at least 12-1/2 percent of the area of the largest phase conductor. If the service conductors are paralleled, the size of the grounded conductor must be based on the total cross-sectional area of the largest set of phase conductors.

When more than one service disconnecting means are located in a listed assembly, only one grounded conductor must be run to and bonded to the service enclosure.

WHEN TWO OR MORE BUILDINGS ARE SUPPLIED FROM A COMMON SERVICE

If two or more buildings are supplied by a single service, the grounded system in each building must have its own ground-ing electrode, connected to the building disconnecting means enclosure. The grounding electrode of each building must have a connection to the grounded service conductor on the load side of the service disconnecting means.

A grounding electrode is not required in separate buildings where only one branch circuit is present, and does not require grounding.

Where two or more buildings are supplied by an ungrounded service, each structure must have a grounding electrode that is connected to the metal enclosure of the building or the structure disconnecting means. The grounding electrode is not required in separate buildings where only one branch circuit is present, and does not require grounding.

DISCONNECTING MEANS ON SAME PREMISES, BUT IN A DIFFERENT BUILDING

When one or more buildings are under the same management, and the disconnects are remotely located, the following conditions must be met:

1. The neutral is to be connected to the grounding electrode at the first building only.
2. Any building with two or more branch circuits has to have a grounding electrode (but only the first building should have its neutral connected to the grounding electrode). The equipment grounding conductor from the first building must be run to this building with the phase conductors, and connected to the grounding electrode just mentioned.
3. The connection of the grounding conductor to the grounding electrode must be made in a junction box located just inside or just outside the building.
4. The grounding conductor, if run underground, must be insulated if livestock are present.

Grounding conductors must be sized according to *Table 250-95*.

GROUNDING FOR SEPARATELY DERIVED SYSTEMS

Separately derived systems are systems that are not taken (derived) from the normal source of power (supplied by the utility company). In almost all cases, the separate system that supplies this power is a transformer. This transformer may be connected to the utility power on its primary side, but its secondary side is a separately derived system nonetheless, and must be grounded as a new source of power, which in fact it is.

The grounding requirements for these systems are as follows:

When separately derived systems must be grounded (a few types do not have to be, mainly because they supply very limited power), the following requirements must be met:

1. Bonding jumpers must be used to connect the equipment grounding conductors of the derived system to the grounded conductor. This connection can be made anywhere between the service disconnecting means and the source, or it can be made at the source of the separately derived system if it has no overcurrent devices or disconnecting means. The bonding jumper cannot be smaller than the grounding electrode conductor (as specified in *Table 250-94*). If the phase conductors are larger than 1100 kcmil copper or 1750 kcmil aluminum, the bonding jumper must be at least 12-1/2 percent of the area of the largest phase conductor. If the service conductors are paralleled, the size of the bonding jumper must be based on the total cross-sectional area of the largest set of phase conductors.

2. A grounding electrode conductor must be used to connect the grounded conductor of the derived system to the grounding electrode. This connection can be made anywhere between the service disconnecting means and the source, or it can be made at the source of the separately derived system if it has no overcurrent devices or disconnecting means. The grounding electrode conductor must be sized as specified in *Table 250-94*. If the phase conductors are larger than 1100 kcmil copper or 1750 kcmil aluminum, the bonding jumper must be at least 12-1/2 percent of the area of the largest phase conductor. If the service conductors are paralleled, the size of the bonding jumper must be based on the total cross-sectional area of the largest set of phase conductors. (Class 1 circuits that are derived from transformers rated no more than 1000 volt-amperes do not require a grounding electrode if the system grounded conductor is bonded to the grounded transformer case.

3. A grounding electrode must be installed next to (or as close as possible to) the service disconnecting means. The grounding electrode can be any of the following:
   a. The nearest structural metal part of the structure that is grounded.
   b. The nearest metal underground water pipe that is grounded.
   c. An electrode (usually No. 4 minimum bare copper wire or reinforcing bar) at least 20 feet long and in direct contact with the earth for its entire length. If reinforcing bar is used, it must be at least 1/2 inch in diameter.
   d. A ring of No. 2 minimum bare copper wire, at least 2-1/2 feet below grade, and encircling the structure (at least 20 feet long).
   e. A ground rod, plate electrode, or made electrode.

CONNECTIONS FOR HIGH-IMPEDANCE GROUND NEUTRALS

Where high-impedance neutral systems are allowed (see *Section 250-5[b], Exception 5*), the following requirements must be observed:

1. The grounding impedance must be installed between the grounding electrode and the neutral conductor.
2. The neutral conductor must be fully insulated.
3. The system neutral can have no other connection to ground, except through the impedance.
4. The neutral conductor between the system source and the grounding impedance can be run in a separate raceway.
5. The connection between the equipment grounding conductors and the grounding impedance (properly called the *equipment bonding jumper*) must be unspliced from the first system disconnecting means to the grounding side of the impedance.
6. The grounding electrode conductor can be connected to the grounded conductor at any point between the grounding impedance and the equipment grounding connections.

GROUNDING ENCLOSURES

All metal enclosures must be grounded, except:

1. Metal boxes for conductors that are added to knob-and-tube, open wiring, or nonmetallic-sheathed cable systems that have no equipment grounding conductor. The runs can be no longer than 25 feet, and guarded against contact with any grounded materials.
2. Short runs of metal enclosures used to protect cable runs.
3. When used with certain Class 1, 2, 3, and fire-protective signaling circuits.

GROUNDING EQUIPMENT

All exposed noncurrent-carrying metal parts of equipment that are likely to become energized must be grounded if any of the following conditions exist:

1. If the equipment is within 8 feet vertically, or 5 feet horizontally, of ground or any grounded metal surface that can be contacted by persons.
2. If the equipment is located in a damp or wet location, unless isolated.
3. If the equipment is in contact with other metal.
4. If the equipment is in a hazardous location.
5. When the enclosure is supplied by a wiring method that supplies an equipment ground (such as metal raceway, metal-clad cable, or metal-sheathed cable).
6. When equipment operates at over 150 volts to ground, except:
   a. Nonservice switches or enclosures that are accessible only to qualified persons.
   b. Insulated electrical heater frames; by special permission only.
   c. Transformers (and other distribution equipment) mounted more than 8 feet above ground or grade.
   d. Listed, double-insulated equipment.

The following types of equipment, regardless of voltage, *must* be grounded:

1. All switchboards, except insulated 2-wire DC switchboards.
2. Generator and motor frames of pipe organs, except when the generator is insulated from its motor and ground.
3. Motor frames.
4. Motor controller enclosures, except for ungrounded portable equipment or lined covers of snap switches.
5. Elevators and cranes.
6. Electric equipment in garages, theatres, and motor picture studios, except pendant lampholders operating at 150 volts or less.

7. Electric signs.
8. Motor picture projection equipment.
9. Equipment supplied by Class 1, 2, 3, or fire-protective signaling circuits, except where specified otherwise.
10. Motor-operated water pumps.
11. The metal parts of cranes, elevators and elevator cars, mobile homes and recreational vehicles, and metal partitions around equipment above 100 volts between conductors.

Except specifically allowed to be otherwise, all cord-and-plug connected equipment must be grounded.

Metal raceways, frames, and other noncurrent-carrying parts of electrical equipment must be kept at least 6 feet away from lightning protection conductors and lightning rods.

GROUNDING METHODS

Equipment grounding connections must be made as follows:

1. For grounded systems—by bonding the equipment grounding conductor

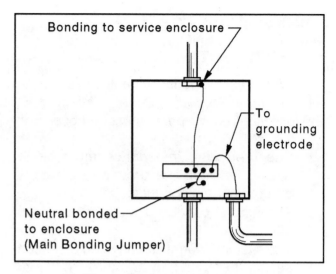

Fig. 7-10 Service bonding.

and grounded service conductor to the grounding electrode conductor. (See Figure 7-10.)

2. For ungrounded systems—by bonding the equipment grounding conductor to the grounding electrode conductor.

Grounding receptacles that replace ungrounded receptacles can be bonded to a grounded water pipe. (See Figure 7-11.)

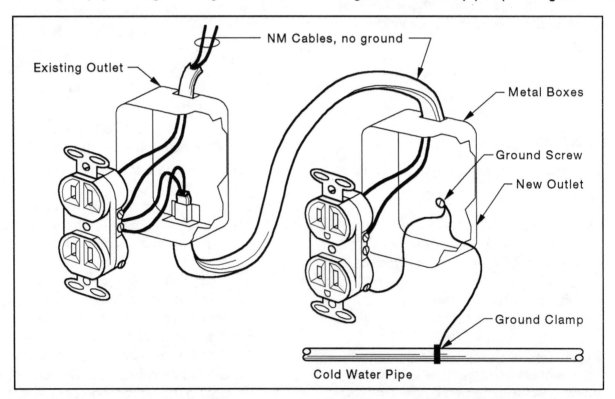

Fig. 7-11 Adding grounding receptacle from ungrounded source.

If there is no ground available in an existing box, it is also allowable to replace a nongrounding type receptacle with a GFI receptacle. Care must be taken, however, that the grounding conductor from the GFI receptacle is not connected to a ground conductor on the load side of the GFI. If this were to be done, it would jeopardize the safety of the installation.

The conductor path to ground from equipment and metal enclosures must:

1. Be permanent and continuous.
2. Have enough capacity for any fault current imposed on it.
3. Have a low enough impedance so as not to limit the voltage to ground.

The earth cannot be used as the only grounding conductor. (In other words, you cannot ground something by simply attaching a wire to its case and pushing it into the ground.)

Only one grounding electrode is allowed at each building. Two or more grounding electrodes that are bonded together are considered to be the same as one grounding electrode.

Metal sheaths of underground service cables can be considered grounded, only because of their contact with the earth and bonding to the underground system. They do not need to be connected to the grounding electrode conductor or grounding electrode. This is also allowed if the cable is installed underground in metal conduit and bonded to the underground system.

Noncurrent-carrying metal parts of equipment, raceways, etc. that require grounding can meet this requirement by being connected to an appropriate equipment grounding conductor.

Electrical equipment is considered grounded if it is secured to grounded metal racks or structures designed to support the equipment. Mounting equipment on the metal frame of a building is *not* considered sufficient for grounding.

See *Chapter 35* of this text for the specific grounding requirements of mobile homes.

Except where specifically permitted, the neutral conductor must *never* be used to ground equipment on the load side of the service disconnect.

If a piece of equipment is connected to more than one electrical system, it must have an appropriate ground connection for each system.

BONDING

Bonding is merely the connecting together of metal parts to form a complete grounding system. Bonding is essential for maintaining the continuity of the grounding system, but it is also important to protect against voltage surges caused by lightning or other fault currents.

The following parts of service equipment must be bonded together:

1. Raceways, cable trays, cable armor, or sheaths.
2. Enclosures, meter fittings, etc.
3. Raceway or armor enclosing a grounding electrode conductor.

One exposed means must be provided for the bonding of other systems (such as telephone or cable TV systems). This can be done by one of the following means:

1. Grounded metal raceway. (Must be exposed.)
2. An exposed grounding electrode conductor.
3. Any other approved means, such as extending a separate grounding conductor from the service enclosure to a terminal strip.

The various components of service equipment must be bonded together by one of the following methods:

1. By bonding the components to the grounded service conductor.
2. By making the connection with rigid or intermediate metal conduit, made up with threaded couplings, or thread-

less couplings and connectors that are made up tight. Standard locknut and bushing connections are *not* sufficient.

3. By connecting bonding jumpers between the various items.
4. By using grounding locknuts or bushings connected to an equipment grounding conductor running between the various items.

Metal raceways and metal-sheathed cables that contain circuits operating at more than 250 volts to ground must be bonded by the same means as for service equipment, as outlined above. (The exception is connection to the grounded service conductor.) If knockouts in enclosures or boxes are field cut, standard methods can be used.

If the uninsulated neutral of a metal service cable is in direct contact with the metal armor or sheath, the sheath or armor is considered grounded without any further connection.

A receptacle's grounding terminal must be bonded to a metal box in which it is installed by one of the following means (see Figure 7-12):

1. By connection with a jumper wire (commonly called a *pigtail*).
2. By receptacle yokes and screws approved for this purpose.
3. By direct metal-to-metal contact between the box and a receptacle's yoke. This applies to surface mounted boxes *only*.
4. By installation in floor boxes listed for this purpose.
5. By connection to an isolated grounding system, where required to eliminate electrical noise in sensitive circuits.

All electrical components that are allowed to act as equipment conductors must be well fitted together to ensure electrical continuity. They *must* be well bonded. These systems include metal raceways, cable trays,

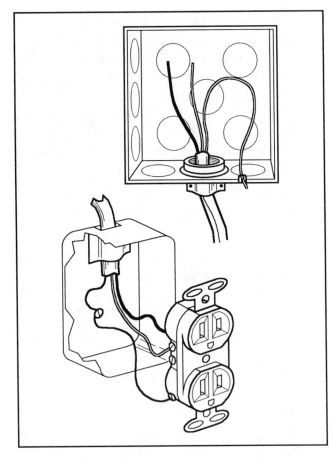

Fig. 7-12 Top box bonded with clip, bottom box bonded with screw.

cable armor, cable sheaths, enclosures, frames, and fittings.

All metal raceways must be made electrically continuous. This requires special care at expansion joints, etc.

Main bonding jumpers connect the grounded service conductor and the service enclosure. This connection should be made according to the instructions supplied by the manufacturer of the service equipment. The bonding jumper cannot be smaller than the grounding electrode conductor (as specified in *Table 250-94*). If the phase conductors are larger than 1100 kcmil copper or 1750 kcmil aluminum, the bonding jumper must be at least 12-1/2 percent of the area of the largest phase conductor. If the service conductors are paralleled, the size of the bonding jumper must be based on the total cross-sectional area of the largest set of phase conductors.

An equipment bonding jumper on the load side of the service must be sized according to *Table 250-95,* based on the largest overcurrent that protects the conductors in the raceways or enclosures.

The bonding jumpers mentioned above can be inside or outside the equipment being bonded. If installed outside, it can be no more than 6 feet long, and must be routed along with the equipment.

Interior metal water piping systems *must* be bonded to one of the following:

1. The service equipment enclosure.
2. The grounded conductor (at the service only).
3. The grounding electrode conductor (unless it is too small).
4. The grounding electrode.

The bonding jumper mentioned above must be sized according to *Table 250-94.*

Other metal piping systems that may be energized must be bonded. The size of the jumper must be based upon *Table 250-95,* calculated upon the rating of the circuit likely to energize the piping.

OTHER REQUIREMENTS

All instruments, etc. in or on switchboards must be grounded.

The grounding conductor for instruments, etc. must be at least No. 12 copper or No. 10 aluminum.

Where the primary voltage of instrument transformers exceeds 300 volts, and the transformers are accessible to unqualified persons, they must be grounded. If inaccessible to unqualified persons, they need not be grounded unless the primary voltage is over 1000 volts.

Instrument transformer cases must be grounded if they are accessible to unqualified persons.

High-voltage systems, where required to be grounded, must meet the same requirements for grounding as low-voltage systems. (See *Sections 250-152* and *250-153* for exceptions.)

SURGE ARRESTERS

Surge arresters are used to send *surge voltages* (unusually large voltages, such as may be caused by a nearby lightning strike) directly to ground, rather than allowing them to affect interior wiring. They are extremely important under some circumstances. (See Figure 7–13.)

Surge arresters (when used) must be connected to all ungrounded conductors.

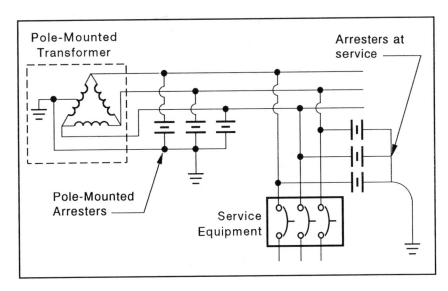

Fig. 7–13 Proper connection of surge arresters.

The rating of a surge arrester (for circuits under 1000 volts) must be equal to or greater than the operating voltage of the circuit it protects.

The rating of a surge arrester (for circuits over 1000 volts) must be at least 125 percent of the operating voltage of the circuit it protects.

Surge arresters can be installed indoors or outdoors. Unless specifically approved for such installations, they must be located where they are not accessible to unqualified persons.

Conductors connecting the surge arrester to the system being protected must be as short as possible.

All surge arresters should be connected according to the manufacturer's directions.

All surge arresters must have a connection to one of the following:

1. The equipment grounding terminal.
2. The grounded service conductor.
3. The grounding electrode conductor.
4. The grounding electrode.

For services of 1000 volts or less, the minimum size for wires connecting surge arresters is No. 14 copper or No. 12 aluminum.

For services of over 1000 volts, the minimum size for wires connecting surge arresters is No. 6 copper or No. 6 aluminum.

When circuits are supplied by 1000 volts or more, and the grounding conductor of a surge arrester protects a transformer that feeds a secondary distribution system, one of the following must be done:

1. If the secondary has a connection to a metal underground water piping system, the surge arrester must have a connection to the secondary neutral.
2. If the secondary does *not* have a connection to a metal underground water piping system, the surge arrester must be connected to the secondary neutral through a spark gap device. (See *Section 280-24[b]* for specifications of the spark gap device.)

Connections other than those mentioned above can be made by special permission.

Chapter Questions

1. Why is it important that return ground paths are clear and of a good quality?

2. Which article in the *NEC*®covers grounding?

3. What does the term *grounded* mean?

4. All AC systems over what voltage must be grounded?

5. Which conductor of a single-phase, 3-wire system must be grounded?

6. What are the most common grounding electrodes?

7. What do we call the conductor between the service disconnect and grounding electrode?

8. Can the armor of Type AC cable be considered a grounding conductor?

9. When must grounding conductors be protected?

10. Are transformers used to supply separately derived systems?

CHAPTER
8

Temporary Wiring

Article 305 of the *NEC*®gives a number of requirements that are more lenient than normal because they apply only to temporary wiring (Christmas tree lighting, for example). These requirements are included in the Code because it would be impractical for every type of temporary wiring to fulfill all of the requirements for permanent wiring (which really means that it would be far too expensive). The idea behind these requirements is to provide for installations that, although they are temporary, are nevertheless safe.

Temporary wiring is allowed for the length of time of construction, demolition, remodeling, etc., or for a maximum of 90 days for Christmas lighting, carnivals, etc.

Temporary wiring may also be used for testing, experimental, and developmental work.

Temporary wiring must be removed as soon as the purpose for its installation is completed.

Temporary services must meet the same requirements as for permanent services.

When the voltage to ground is not greater than 150 volts, feeders can be run as open conductors. In such cases, they must be supported on insulators every 10 feet or less.

When the voltage to ground is not greater than 150 volts, branch circuits can be run as open conductors. In such cases, they must be supported on insulators every 10 feet or less. They may *not* be laid on floors.

Full overcurrent protection must be provided for temporary wiring, the same as for permanent wiring.

Multiconductor cords and cables can be used for temporary branch circuits, but the cord or cable must be listed for hard usage.

All receptacles must be of the grounding type.

Unless run in metal raceway or metal-covered cable, all branch circuits must have an equipment grounding conductor.

Receptacles cannot be connected to the same circuits that supply temporary lighting.

All lamps for general lighting must be equipped with a guard or housed in a fixture.

Metal-cased lighting sockets cannot be used unless the sockets are grounded.

Boxes are not required for splices in multiconductor cords or cables on construction sites.

A box, fitting, etc. with a bushed hole for each conductor must be used where changing to or from raceway systems and open wiring.

Flexible cords and cables must be protected from damage. Pinch points, such as passing cords and cables through doorways, must be protected.

All single-phase, 125-volt, 15- and 20-ampere receptacles on construction sites must be ground-fault protected.

A testing system must be enforced at each construction site where temporary wiring is used. The following tests must be performed on all cord sets, receptacles, and cord-and-plug connected equipment:

1. All equipment grounding conductors must be checked for continuity.
2. Each receptacle and plug must be checked to verify that the equipment grounding conductor is correctly attached.

The above tests must be done:

1. Before their first use at the construction site.
2. If there is evidence of damage.
3. After repairs, before being returned to service.
4. No more than once every three months.

The tests must be recorded, and a record of the tests must be made available for review by the local inspector.

Fencing or other suitable means of guarding must be installed around temporary wiring operating at over 600 volts.

Chapter Questions

1. Can temporary wiring be used for testing processes?

2. What is the maximum length of time temporary wiring can be used?

3. Must boxes be used for splices in temporary wiring?

4. Which lamps on temporary wiring systems must be provided with guards?

5. Can lighting and receptacles be installed on the same temporary power circuits?

6. Are there relaxed requirements for temporary services?

7. Do temporary circuits in AC cables need a separate grounding conductor?

8. If cords are used for temporary wiring, what type must they be?

9. Can open conductors be used for 120/208-volt, 3-phase, 4-wire systems?

10. Can open wiring be used for 277/480-volt, 3-phase, 4-wire temporary systems?

PART
2

RACEWAY SYSTEMS

CHAPTER
9

Conduit Systems

Enclosing conductors in conduit remains the best method of providing mechanical protection to the conductors. The only problem with conduit is its cost. Not only are conduits frequently expensive to purchase (material cost), but installation is rather labor-intensive. Conduits must be cut and bent to the exact size and shape required for any installation. In addition, most types of conduits require expensive benders, thus adding substantially to their installed cost.

To remedy this problem, a number of different types of conduit have been developed that have reduced costs somewhat. For example, EMT (which technically is not conduit but "tubing") can be easily cut, and can be bent with an inexpensive hand bender. And because it has thinner walls than rigid conduit, it is less expensive in material costs as well. PVC conduits are also less expensive in material costs than heavy walled conduits, and they also are coupled with glued joints rather than pipe fittings.

In this chapter, we will cover the various conduit systems of the *NEC*®. Remember, however, that these requirements are dependent upon good workmanship. (Remember what I said in *Chapter 1* about doing installations in a "neat and workmanlike manner.") The vast majority of defects in conduit systems are not due to violations of requirements (such as support spacings), but rather to poor workmanship.

ELECTRICAL METALLIC TUBING

Electrical metallic tubing (commonly called "EMT" or "Thin-wall") is the most commonly used type of conduit. (As I said earlier, EMT is not really considered conduit in a technical sense, but for all practical purposes it *is* conduit.) Because it is rather inexpensive both to purchase and to install, and because its installation requires no expensive tools, it is chosen in almost all circumstances where it is allowed.

The requirements of EMT are given in *Article 348* of the *NEC*®. They are as follows:

Where Permitted

EMT May be installed in all locations *except*:

1. Where it will be vulnerable to severe damage.
2. In concrete or cinder fill. (It may be installed in these locations if it is enclosed on all sides by 2 or more inches of concrete, or if the the tubing is 18 or more inches below the fill.
3. In any hazardous location, unless specifically permitted. (See *Sections 502-4, 503-3,* and *504-20.*)

4. In concrete, earth, or corrosive areas, except when protected to the satisfaction of the local authorities.

Installation Methods

EMT may not be threaded.

Raintight (compression) fittings must be used in wet (outdoor) locations.

Bends must not crimp or flatten the tubing. Use only benders designed for the purpose.

No more than 360° of bend (equivalent to four quarter bends) are allowed between pulling points.

All cut ends must be reamed.

EMT must be strapped (supported) within 3 feet of every box, cabinet, or fitting (condulet type), and every 10 feet thereafter.

If support is not possible within 3 feet of a box, cabinet, or fitting, an unbroken piece of EMT ("unbroken" meaning with no couplings in the section) may run 5 feet before it is supported.

RIGID METAL CONDUIT

Rigid metal conduit is perhaps the strongest and most reliable type of conduit system available. It has been used for many years with excellent results. In fact, unless the conduit is installed in an unprotected area where it could rust, it can last for many decades.

The requirements for rigid metal conduit (often called "GRC" [galvanized rigid conduit] or "Heavy-wall") are found in *Article 346* of the *NEC*®, and are as follows:

Uses and Locations

Rigid metal conduit is acceptable for all uses in all locations, *with the following restrictions:*

1. When installed in contact with earth or concrete, or in corrosive areas, it must be suitably protected.

2. When installed in or under cinder fill, it must be enclosed on all sides by 2 or more inches of concrete, submerged 18 or more inches below the fill, or suitably protected.

Installation Methods

It is recommended (*not* required) that contact between rigid metal conduit and different types of metals be avoided, to prevent galvanic action.

All cut ends must be reamed.

Bushings must be used to protect wires every time rigid metal conduit enters a box, fitting, or enclosure. (There are a few types of boxes, fittings, and enclosures that have built-in protection for wires. Obviously, in these cases bushings are not required.)

Threadless fittings must be suitable for the area in which they are installed. (Concretetight fittings must be used in concrete, and raintight [compression] fittings must be used in wet locations.)

Running threads may not be used where the conduit connects to couplings.

Conduit may not be flattened when bent. Use benders designed for the purpose.

No more than 360° of bend is allowed between pulling points.

Conduit must be supported within 3 feet of every box, cabinet, or fitting (condulet type), and every 10 feet thereafter.

If only threaded couplings are used and proper supports are used, 1-inch conduit may be supported every 12 feet; 1-1/4-inch and 1-1/2-inch conduit may be supported every 14 feet; 2-inch and 2-1/2-inch conduit may be supported every 16 feet; and conduit 3 inches and larger may be supported every 20 feet.

Runs of conduit from industrial machinery that uses only threaded couplings, and is supported both top and bottom, may be supported only every 20 feet.

INTERMEDIATE METAL CONDUIT

Intermediate metal conduit (commonly called "IMC") is a modification of rigid metal conduit. The difference between the two is that IMC is made of somewhat thinner and stiffer metal.

The requirements for IMC are covered in *Article 345* of the *NEC*®, and are as follows:

Uses and Locations

Intermediate metal conduit is acceptable for all uses in all locations, *with the following restrictions:*

1. When installed in contact with earth or concrete, or in corrosive areas, it must be suitably protected.
2. When installed in or under cinder fill, it must be enclosed on all sides by 2 or more inches of concrete, submerged 18 or more inches below the fill, and suitably protected.

Installation Methods

It is recommended (*not* required) that contact between intermediate metal conduit and different types of metals be avoided, to prevent galvanic action.

All cut ends must be reamed.

Bushings must be used to protect wires every time intermediate metal conduit enters a box, fitting, or enclosure. (There are a few types of boxes, fittings, and enclosures that have built-in protection for wires. Obviously, in these cases bushings are not required.)

Threadless fittings must be suitable for the area in which they are installed. (Concretetight fittings must be used in concrete, and raintight [compression] fittings must be used in wet locations.)

Running threads may not be used where the conduit connects to couplings.

Conduit may not be flattened when bent. Use benders designed for the purpose.

No more than 360° of bend is allowed between pulling points.

Conduit must be supported within 3 feet of every box, cabinet, or fitting (condulet type), and every 10 feet thereafter.

If only threaded couplings are used and proper supports are used, 1-inch conduit may be supported every 12 feet; 1-1/4-inch and 1-1/2-inch conduit may be supported every 14 feet; 2-inch and 2-1/2-inch conduit may be supported every 16 feet; and conduit 3 inches and larger may be supported every 20 feet.

Runs of conduit from industrial machinery that uses only threaded couplings, and is supported both top and bottom, may be supported only every 20 feet.

RIGID NONMETALLIC CONDUIT

Rigid nonmetallic conduit (usually Schedule 40 PVC) is excellent for underground runs, since it is nonmetallic and cannot rust. It is also easy to install (glued joints), and is rather inexpensive.

The support of this type of conduit is an important factor, since it tends to bend when it becomes hot. Runs of PVC conduit tend to sag or droop when installed out of doors (underground runs obviously excepted). The heat of summer is enough to cause this. Therefore, these runs must be supported at closer intervals than is required for metal conduits.

The requirements for rigid nonmetallic conduit are found in *Article 347* of the *NEC*®.

Uses and Locations

Rigid nonmetallic conduit is acceptable for all uses and locations, *except:*

1. It may not be used in hazardous locations, except where specifically permitted. (See *Sections 504-20, 514-8, 515-5,* and *501-4[b]* for exceptions.)
2. It may not be used to support lighting fixtures.
3. It may not be used in locations where it will be subjected to physical damage.

4. Rigid nonmetallic conduit may be used in theatres and similar locations only when it is encased in 2 or more inches of concrete.

5. It may not be used in locations where the temperature exceeds the conduit's rated temperature. (The temperature rating of the conduit is usually stamped on the conduit at 10-foot intervals.)

Installation Methods

All cut ends must be trimmed to remove rough edges.

Rigid nonmetallic conduit must be support within 3 feet of every conduit termination. Thereafter, 1/2-inch, 3/4-inch, and 1-inch rigid nonmetallic conduit must be supported every 3 feet; 1-1/4-inch, 1-1/2-inch, and 2-inch conduit must be supported every 5 feet; 2-1/2-inch and 3-inch conduit must be supported every 6 feet; 3-1/2-inch, 4-inch, and 5-inch conduit must be supported every 7 feet; and 6-inch conduit must be supported every 8 feet.

Expansion joints must be installed in long runs of rigid nonmetallic conduit. (Exactly what length of run requires an expansion joint is not specified in the Code, but *Table 10* in *Chapter 9* of the Code shows the expansion characteristics of PVC conduit.)

Bushings must be used in terminating rigid nonmetallic conduit, unless the box or termination fitting provides protection. (Virtually all do.)

Bends in rigid nonmetallic conduit must be made with benders designed for the purpose, and must not flatten or crimp the conduit.

No more than 360° of bend is allowed between pulling points.

FLEXIBLE METAL CONDUIT

Flexible metal conduit is very easy to install, since it requires no mechanical bending. It is

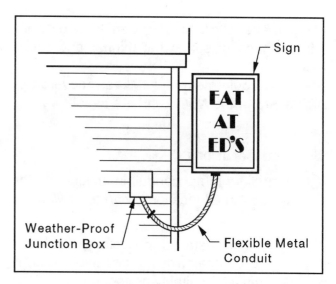

Fig. 9-1 Outdoor sign connected with flexible metal conduit.

installed in almost the same manner as cable, which makes installation very inexpensive.

The requirements for flexible metal conduit (commonly called "Greenfield" or "Flex") are found in *Article 350* of the *NEC®*, and are as follows:

Uses and Locations

Flexible metal conduit may *not* be used in the following locations:

1. In wet locations, unless lead-covered conductors are used, or approved for the specific installation. (See Figure 9-1.)
2. In hoistways, except as specifically permitted. (See *Section 620-21*.)
3. In storage battery rooms.
4. In hazardous locations, except as specifically permitted. (See *Sections 501-4[b]* and *504-20*.)
5. Underground.
6. Embedded in concrete or aggregate.

Three-eighths inch flexible metal conduit may be used only for underplaster extensions, enclosing motor leads, lighting fixture whips, on elevators (only when less than 6 feet long), and as factory manufactured systems.

Installation Methods

Flexible metal conduit must be supported within 12 inches of every box, cabinet, or fitting, and every 4-1/2 feet thereafter, *except* when used as fixture whips, fished into place, or in lengths of less than 3 feet at terminals where flexibility is necessary.

Flexible metal conduit can be used as a grounding conductor only if the run is less than 6 feet long, uses fittings approved for grounding, and the circuit is protected at 20 amperes or less.

No more than 360° of bend is allowed between pulling points.

Angle connectors cannot be used on concealed raceways.

FLEXIBLE METALLIC TUBING

Flexible metallic tubing is used primarily in plenums and airways. Its requirements are similar to those for flexible metal conduit, but for most installations flexible metal conduit is a better choice.

The requirements for flexible metallic tubing are covered in *Article 349* of the *NEC®*. They are as follows:

Where Permitted

For dry, accessible, protected locations; and only for branch circuits less than 1000 volts, in lengths of 6 feet or less.

Where Not Permitted

In hoistways.

In storage battery locations.

In hazardous locations (except where specifically permitted).

Underground.

Embedded in concrete or aggregate.

In areas where it might be subjected to physical damage.

Installation Methods

Three-eighths inch flexible metallic tubing may be used only in environmental air ducts, plenums, or spaces; or as lighting fixture whips.

Three-quarter inch flexible metallic tubing is the largest size permitted.

Flexible metallic tubing can be used as a grounding conductor only if the run is less than 6 feet long, uses fittings approved for grounding, and the circuit is protected at 20 amperes or less.

The minimum bend radii for fixed bends of flexible metallic tubing are 3-1/2 inches for 3/8-inch tubing; 4 inches for 1/2-inch tubing; and 5 inches for 3/4-inch tubing.

The minimum bend radii for flexing bends of flexible metallic tubing are 10 inches for 3/8-inch tubing; 12-1/2 inches for 1/2-inch tubing; and 17-1/2 inches for 3/4-inch tubing.

LIQUIDTIGHT FLEXIBLE METAL CONDUIT

("Seal-tite," *Article 351*)

Where Permitted

Where flexibility is required.

Where conditions in the area of installation require liquidtight or vaportight protection; either exposed, concealed, or underground.

In hazardous locations (but *not* Class 1, Division 1) and floating buildings where flexibility is necessary. (See *Sections 501-4[b], 502-4, 503-3, 504-20,* and *553-7[b]*.)

Where Not Permitted

Where it may be subjected to physical damage.

When the operating temperature might be higher than the rating of the liquidtight conduit.

Installation Methods

Three-eighths inch liquidtight flexible metal conduit may be used only in environmental air ducts, plenums, or spaces; or as lighting fixture whips. A minimum of 1/2-inch must be used for all other installations.

Liquidtight flexible metal conduit must be supported within 12 inches of every box, cabinet, or fitting, and every 4-1/2 feet thereafter, *except* when used as fixture whips, fished into place, or in lengths of less than 3 feet at terminals where flexibility is necessary.

Liquidtight flexible metal conduit can be used as a grounding conductor only if the run is less than 6 feet long, uses fittings approved for grounding, and the circuit is protected at 20 amperes or less for 3/8-inch and 1/2-inch liquidtight; 60 amps or less for 3/4-inch, 1-inch, and 1-1/4-inch liquidtight.

No more than 360° of bend is allowed between pull points.

Angle connectors are not to be used in concealed installations.

LIQUIDTIGHT FLEXIBLE NONMETALLIC CONDUIT

("Seal-tite," *Article 351*)

Where Permitted

Where flexibility is required.

Where conditions in the area of installation require liquidtight or vaportight protection; either exposed, concealed, or underground.

Outdoors, when marked as suitable for that use.

Directly buried, when marked as suitable for that use.

Where Not Permitted

Where it may be subjected to physical damage.

When the operating temperature might be higher than the rating of the liquidtight conduit.

In hazardous locations.

When the voltage of the enclosed conductors exceeds 600 volts.

Installation Methods

Lengths of more than 6 feet are not allowed, unless approved (by the local inspector) for the specific installation.

Three-eighths inch liquid flexible nonmetallic conduit may be used only for enclosing motor leads.

All fittings must be identified for use with liquidtight flexible nonmetallic conduit.

Equipment grounding conductors must be used.

Exterior equipment grounding conductors may not be longer than 6 feet, and must be routed with the conduit.

All boxes and fittings must be bonded.

Chapter Questions

1. Does the *NEC*®require a separate grounding conductor to be run in EMT?

2. Why can't EMT be installed in cinder fill?

3. Why are closely spaced supports needed for PVC conduit?

4. What must be done to the cut ends of rigid metal conduit?

5. When can running threads not be used on rigid metal conduit?

6. How far apart must supports be placed on 3-inch rigid metal conduit that has threaded couplings?

7. How many degrees of bend are allowed in runs of conduit?

8. What must be done to rigid nonmetallic conduit if it is to be used in theatres?

9. Can flexible metal conduit be used in wet locations?

10. Where can 3/8-inch liquidtight metallic conduit be used?

11. Under normal circumstances, what is the maximum length of run for liquidtight nonmetallic conduit?

CHAPTER
10

Raceways and Wireways

Raceways and wireways are various types of enclosures, through which are run electrical conductors. There are a number of different types of raceways and wireways, covered under several different articles in the *NEC®*. They are as follows:

SURFACE METAL RACEWAYS

("Wiremold," *Article 352*)

Surface metal raceway is the metal wiring channel with removable covers that is commonly called "Wiremold," although it may or may not be made by the Wiremold Company. It is especially useful for adding wiring once a building finish has been completed. It is run along the surface of buildings, the conductors installed, and the covers snapped on.

Where Permitted

In dry locations. (See Figures 10-1 and 10-2.)

As underplaster extensions, when identified as suitable for such use.

Under raised floors. (See *Section 645-5[d][2]*.)

Where Not Permitted

Where surface metal raceway might be subjected to severe physical damage.

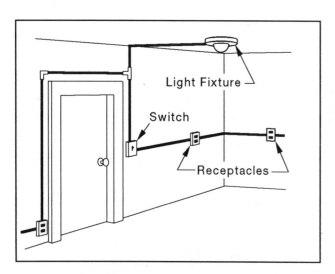

Fig. 10-1 Typical run of surface raceway.

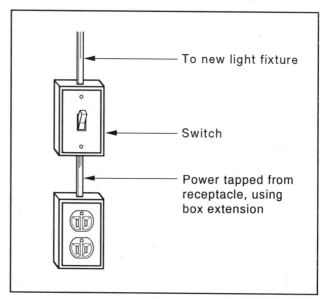

Fig. 10-2 Taking power from existing outlet, using surface raceway.

It may not be used to enclose conductors that have a voltage of 300 or higher between conductors, unless the metal is at least 0.040 inches thick.

In areas where hazardous vapors are present.

In concealed locations, except for under-plaster extensions and under raised floors.

In hazardous locations, except when enclosing conductors in Class 1, Division 2, locations that do not carry enough power to ignite specific atmospheres. (See *Section 504-4[b],* Exception.)

Installation Methods

Only unbroken pieces of surface metal raceway may be extended through floors, walls, or partitions.

When divided raceways are used (usually one section for power wiring, and one or more sections for data or communications wiring), the positions of the sections relative to one another (such as "data in the left-hand section and power in the right-hand section") must be maintained throughout the installation. (This avoids confusion.)

Splices or taps are allowed only in surface raceways that are accessible after installation and have a removable cover. These splices may not occupy more than 75 percent of the cross-sectional area of the raceway at any point.

Splices for surface raceways that do not have removable covers are allowed only in junction boxes.

All sections of surface metal raceway must be mechanically joined so that the conductors will not be subjected to abrasion, and electrically connected so that they will all be properly grounded.

Any nonmetallic covers, etc. used with surface metal raceways must be identified as suitable for that use.

SURFACE NONMETALLIC RACEWAYS

(*Article 352*)

Surface nonmetallic raceway is similar to surface metal raceway; however, the Code omits a number of details and requirements for its installation. For instance, can it be extended through walls, floors, and partitions? Are splices and taps allowed? May it be used under raised floors? It is my opinion that the rules for surface metal raceway covering these questions should also be applied to surface nonmetallic raceway (with the exception that only metal raceways should be permitted to go through a partition). However, remember that the "authority having jurisdiction" (the local inspector) has the final say in matters of interpretation; therefore, consult with the inspector before going ahead with any questionable installations.

Where Permitted

In dry locations.

Where Not Permitted

Where surface nonmetallic raceway might be subjected to severe physical damage.

It may not be used to enclose conductors that have a voltage of 300 or higher between conductors, unless specifically listed for a higher voltage.

In concealed locations.

In any hazardous locations, except when enclosing conductors in Class 1, Division 2, locations that do not carry enough power to ignite specific atmospheres. (See *Section 504-4[b],* Exception.)

Surface nonmetallic raceway cannot be used to enclose conductors that have temperature limitations higher than those of the surface nonmetallic raceway.

Installation Methods

Only unbroken pieces of surface nonmetallic raceway may be extended through floors, walls, or partitions.

When divided raceways are used (usually one section for power wiring, and one or more sections for data or communications wiring), the positions of the sections relative to one another (such as "data in the left-hand section and power in the right-hand section") must be maintained throughout the installation. (This avoids confusion.)

All sections of surface nonmetallic raceway must be mechanically joined so that the conductors will not be subjected to abrasion.

MULTIOUTLET ASSEMBLIES

(*Article 353*)

Multioutlet assemblies are raceways that have receptacle outlets installed in them. Multioutlet assemblies are most commonly known as "Plug-mold."

Where Permitted

Exposed, in dry locations.

Where Not Permitted

Concealed. (The back and sides may be recessed in the building's finish, but not the front.)

Multioutlet assemblies cannot be used where they may be subjected to physical damage.

They may not be used to enclose conductors that have a voltage of 300 or higher between conductors, unless the metal is at least 0.040 inches thick.

In areas where hazardous vapors are present.

Multioutlet assemblies may not be used in hoistways.

In hazardous locations, except when enclosing conductors in Class 1, Division 2, locations that do not carry enough power to ignite specific atmospheres. (See *Section 504-4[b]*, Exception.)

Installation Methods

It is allowable to install metal multioutlet assemblies through partitions, as long as the

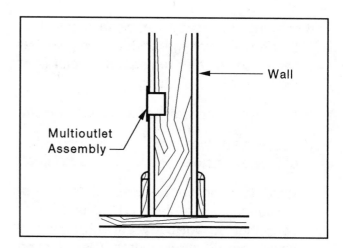

Fig. 10-3 Multioutlet assembly recessed in wall.

cap covering all exposed portions can be removed, and no outlet is located within the partition. (See Figures 10-3 through 10-7.)

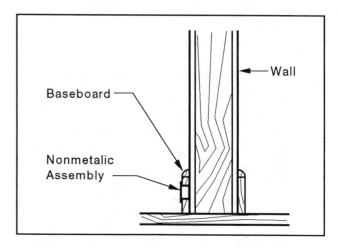

Fig. 10-4   Nonmetallic assembly recessed in baseboard.

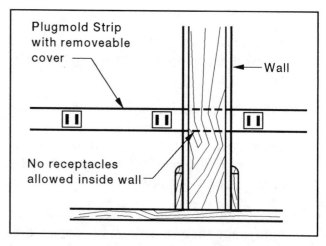

Fig. 10-5 Multioutlet strip passing through wall.

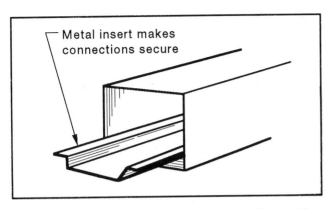

Fig. 10-6 Surface raceway, showing metal mounting insert.

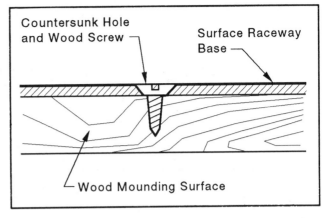

Fig. 10-7 Base of surface raceway, showing mounting method.

UNDERFLOOR RACEWAYS

(*Article 354*)

Where Permitted

Underneath concrete floors. (See Figure 10-8.)

Laid flush with finished floors in offices, when covered by a linoleum (or equivalent) floor finish. (See Figure 10-9.)

Where Not Permitted

Underfloor raceways may not be used where they will be subjected to corrosive vapors.

In hazardous locations, except when enclosing conductors in Class 1, Division 2, locations that do not carry enough power to ignite specific atmospheres. (See *Section 504-4[b]*, Exception.)

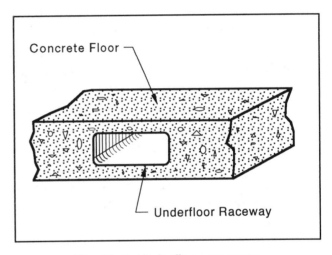

Fig. 10-8 Underfloor raceway.

Installation Methods

Half-round or flat-top raceways that are 4 inches or less wide must be covered with

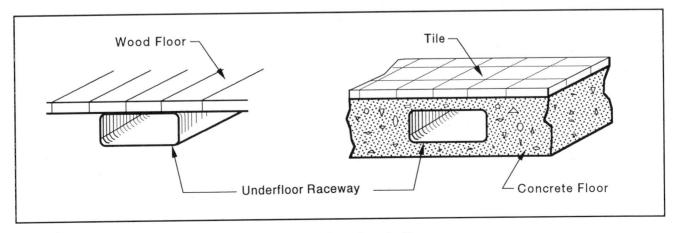

Fig. 10-9 Flush mounting of underfloor raceways.

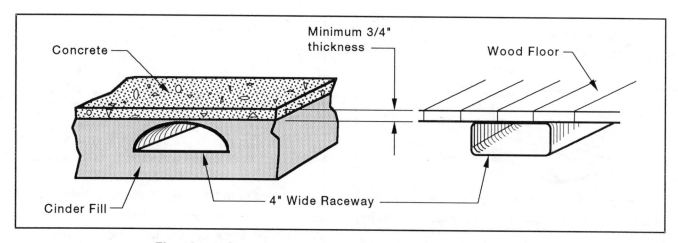

Fig. 10-10 Coverage requirements for underfloor raceways.

at least 3/4 inch of concrete or wood. (See Figure 10-10.)

Flat-top raceways between 4 and 8 inches wide and with at least a 1-inch space between other raceways must be covered with at least 1 inch of concrete. When these raceways are less than 1 inch apart, they must be covered with at least 1-1/2 inches of concrete. (See Figure 10-11.)

Only trench-type raceways with removable covers can be installed flush with the finished floor.

In offices, raceways no more than 4 inches wide with flat covers can be installed flush with the finished floor, but only if they are covered with at least 1/16 inch of linoleum or the equivalent. Up to three such raceways can be installed parallel to each other, but in such instances they must be joined together to form a rigid assembly.

All splices and taps must be made in junction boxes, *not* in the raceway itself. However, when using trench ducts (which have flat tops finished flush with the floor) with removable covers accessible after installation, the splices may not occupy more than 75 percent of the cross-sectional area of the raceway. (See Figure 10-12.)

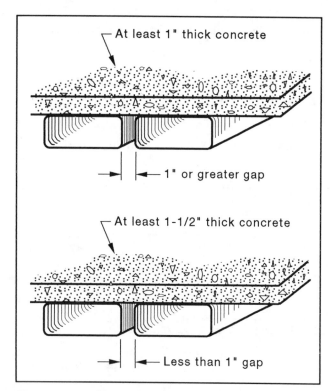

Fig. 10-11 Spacing requirements for multiple underfloor raceways

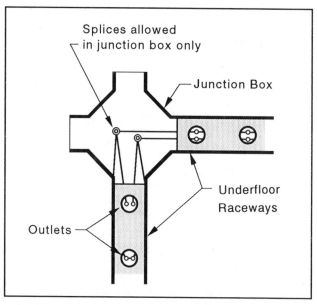

Fig. 10-12 Splicing in underfloor raceways.

"Loop wiring" (when conductors are not spliced but rather looped at each outlet, then put under a terminal screw with insulation removed but the conductor unbroken) is not considered to be a splice.

When any outlet is abandoned or removed, all conductors serving the outlet must be removed from the raceway. Looped conductors may *not* be reinsulated and put back in the raceway; they must be removed.

Underfloor raceways must be laid in straight lines, with no offsets, kicks, or bends. They must be firmly secured during concrete pouring, so that the straight-line alignment of the raceway will not be disturbed.

A marker must be placed near the end of every raceway, so that the last insert may be located easily.

All open ends of raceway must be closed.

All junction boxes must be set level with the floor and sealed so that no water may enter. Only metal boxes may be used with metal raceways, and they must be electrically bonded to the raceway.

All inserts must be set level with the floor and sealed so that no concrete may enter during pouring. Only metal inserts may be used with metal raceways, and they must be electrically bonded to the raceway.

When fiber raceway is used, the inserts used with it must be secured to the raceway before concrete is poured. If inserts are to be added to fiber raceway after concrete is set, they must be screwed into the raceway. When this is done, proper tools must be used, so that the tool does not protrude into the raceway and damage the conductors. Also, all debris must be removed from the raceway.

When underfloor raceways are to be connected to wall outlets or cabinets, one of the following methods must be used:

1. Rigid metal conduit.
2. Intermediate metal conduit.
3. Electrical metallic tubing.
4. Approved raceway fittings.
5. Underfloor raceway encased in concrete.
6. Flexible metal conduit.

If metal underfloor raceway has provision for an equipment grounding conductor connection, connection between the raceway and wall outlets or cabinets may be made by any of the following methods, in addition to those mentioned above:

1. Rigid nonmetallic conduit.
2. Electrical nonmetallic tubing.
3. Liquidtight flexible nonmetallic conduit.

CELLULAR METAL FLOOR RACEWAYS

(Article 356)

Cellular metal floor raceways use the open spaces of cellular floors to enclose wiring. Fittings are used to go between the cells (enclosed tubular spaces in the floor) and the area being served. These fittings extend through the concrete that is poured around the cells, making the finished floor. Header ducts are transverse raceways that provide access to the various cells. (See Figures 10-13 through 10-15.)

Where Not Permitted

Since these raceways are part of the structure of a building, the question is not "Where are these raceways allowed?", but rather "In

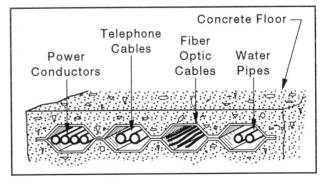

Fig. 10-13 Usage of cellular metal floor raceways.

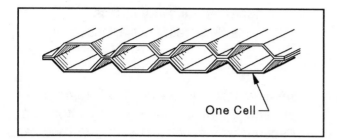

Fig. 10-14 Typical cellular metal floor raceway.

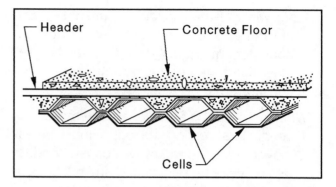

Fig. 10-15 Use of header duct.

what locations can these raceways actually have conductors installed, and be used?"

Where corrosive vapors are present.

In hazardous locations, except when enclosing conductors in Class 1, Division 2, locations that do not carry enough power to ignite specific atmospheres. (See *Section 504-4[b],* Exception.)

In commercial garages cellular metal floor raceways may only be used for supplying power to outlets below a floor, never above a floor.

Installation Methods

Conductors installed in these raceways must be no larger than No. 1/0 AWG, except when special permission is given by the authority having jurisdiction.

Splices are not permitted, except in header ducts or in junction boxes.

"Loop wiring" (in which conductors are not spliced but rather looped at each outlet, then put under a terminal screw with insulation removed but the conductor unbroken) is not considered to be a splice.

When any outlet is abandoned or removed, all conductors serving the outlet must be removed from the raceway. Looped conductors may *not* be reinsulated and put back in the raceway; they must be removed.

A sufficient number of markers must be installed, so that cells may be located in the future.

All junction boxes must be set level with the floor and sealed so that no concrete or water may enter. Only metal boxes may be used, and they must be electrically bonded to the raceway.

All inserts must be set level with the floor so that no concrete may enter during the pouring of the floor. Only metal inserts may be used, and they must be electrically bonded to the raceway.

When cellular metal floor raceways are to be connected to wall outlets or cabinets, one of the following methods must be used:

1. Rigid metal conduit.
2. Intermediate metal conduit.
3. Electrical metallic tubing.
4. Approved raceway fittings.
5. Cellular metal floor raceway encased in concrete.
6. Flexible metal conduit.

If the raceway has provision for an equipment grounding conductor connection, connection between the raceway and wall outlets or cabinets may be made by any of the following methods, in addition to those mentioned above:

1. Rigid nonmetallic conduit.
2. Electrical nonmetallic tubing.
3. Liquidtight flexible nonmetallic conduit.

All surfaces must be kept free of burrs or rough edges that can damage the conductors. Smooth fitting or bushings must be used when conductors pass through the walls of the raceway.

CELLULAR CONCRETE FLOOR RACEWAYS

(Article 358)

Cellular concrete floor raceways use the open spaces of precast concrete floors to enclose wiring. Metal fittings are used to go between the cells (enclosed tubular spaces in the floor) and the area being served. Headers are transverse metal raceways that provide access to the various cells. (See Figure 10–16.)

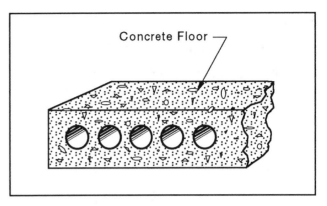

Fig. 10-16 **Cellular concrete floor raceways.**

Where Not Permitted

As with cellular metal floor raceways, the question here is which of the building's raceways are allowed to be used.

Where corrosive vapors are present.

In hazardous locations, except when enclosing conductors in Class 1, Division 2, locations that do not carry enough power to ignite specific atmospheres. (See *Section 504-4[b],* Exception.)

In commercial garages cellular concrete floor raceways may be used only for supplying power to outlets below a floor, never above a floor.

Installation Methods

Conductors installed in these raceways must be no larger than No. 1/0 AWG, except when special permission is given by the authority having jurisdiction.

Splices are not permitted, except in header ducts or in junction boxes.

"Loop wiring" (in which conductors are not spliced but rather looped at each outlet, then put under a terminal screw with insulation removed but the conductor unbroken) is not considered to be a splice.

When any outlet is abandoned or removed, all conductors serving the outlet must be removed from the raceway. Looped conductors may *not* be reinsulated and put back in the raceway; they must be removed.

A sufficient number of markers must be installed so that cells may be located in the future.

All junction boxes must be set level with the floor and sealed so that no concrete or water may enter. Only metal boxes may be used, and they must be electrically bonded to the raceway.

All inserts must be set level with the floor and sealed so that no concrete or water may enter during the pouring of the floor. Only metal inserts may be used, and they must be terminated in grounding type receptacles.

A separate grounding conductor must be used to connect the receptacles to a ground connection at the header.

If inserts are added to these raceways, or whenever the cell walls must be cut, proper tools must be used so that the tool does not protrude into the raceway and damage the conductors. Also, all debris must be removed from the raceway.

When cellular concrete floor raceways are to be connected to wall outlets or cabinets, metal raceways and approved fittings must be used.

Headers must be installed perpendicularly, in straight lines. The ends of the header must be sealed with metal covers, and the header must be secured to the *top* of the precast concrete floor. Headers must be bonded to the distribution center enclosure, and must be electrically continuous for their entire length.

WIREWAYS

(Gutters, troughs, *Article 362*)

Wireways are extended rectangular raceways that are frequently used to connect a number of pieces of electrical equipment together. They are quite popular, since they allow a great deal of freedom in wiring and make it very easy to change wiring after it is initially installed.

Where Permitted

Only in exposed locations, except when used to enclose wiring for sound recording equipment. In these cases, the wireway must be run in straight lines; the covers must be accessible; and no sharp or rough edges are permitted. (See Figure 10–17.)

Where Not Permitted

Where it may be subjected to corrosive vapors.

Where it may be subjected to physical damage.

In hazardous locations, except when enclosing conductors in Class 1, Division 2, and Class 2, Division 2, locations that do not carry enough power to ignite specific atmospheres. (See *Section 504-4[b]*, Exception.)

Installation Methods

Splices or taps are allowed in accessible locations, but may not occupy more than 75 percent of the wireway's cross-sectional area.

Wireways must have supports placed no more than 5 feet apart, except for pieces longer than 5 feet that are supported at each end or joint. (Check to see if the wireway is specifically listed for support intervals.) In no case can the supports be placed more than 10 feet apart.

Wireways that run vertically can have supports placed up to 15 feet apart, as long as there is no more than one joint between supports.

Unbroken sections of wireway are permitted to pass through walls. (See Figure 10–18.)

All open ends of wireway must be covered.

Wireways must be grounded. (See *Article 250*.)

Where grounding connections are made to wireways, any paint or coating must be removed from the area of the connection.

Wire Fills

Wireways are not allowed to contain more than 30 current-carrying conductors (neu-

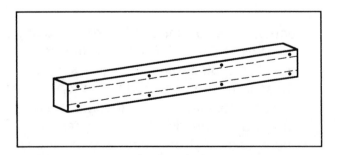

Fig. 10-17 Typical wireway.

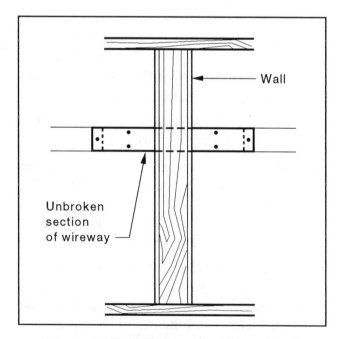

Fig. 10-18 Wireway passing through wall.

tral and grounding conductors or control wiring conductors are not counted) at any one cross-sectional area, unless using the correction factors of *Article 310,* Note 8(a) of Notes to the Ampacity Tables.

The 30 wire limitation is not required in theatres and similar locations.

The total of the cross-sectional areas of all conductors in the wireway may not be greater than 20 percent of the cross-sectional area of the wireway.

Wireways for elevators are not subject to the two preceding requirements.

AUXILIARY GUTTERS

(Article 374)

Auxiliary gutters are wireways that supplement wiring at meter centers, switchboards, etc. (See Figures 10–19 and 10–20.)

Where Permitted

Auxiliary gutters can enclose either conductors or busbars.

Auxiliary gutters may *not* contain switches, appliances, overcurrent devices, or similar items.

Installation Methods

Auxiliary gutters may not extend more than 30 feet from the equipment it serves, except when used for elevators.

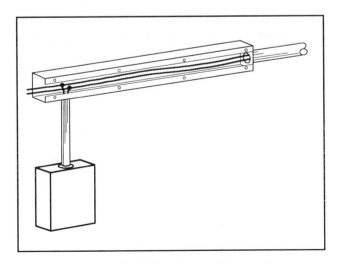

Fig. 10–20 Splicing in auxiliary gutter.

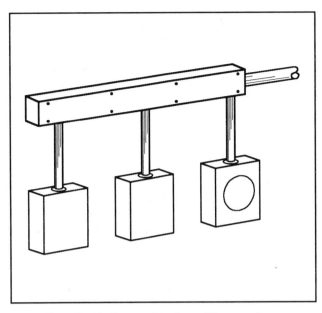

Fig. 10–19 Purpose of auxiliary gutters.

Gutters must be supported at least once every 5 feet.

Splices or taps are allowed in accessible locations, but may not occupy more than 75 percent of the gutter's cross-sectional area.

All taps in gutters must have overcurrent protection (see *Section 240-21*). They must also be marked, showing the circuit or equipment they supply. (See Figure 10–21.)

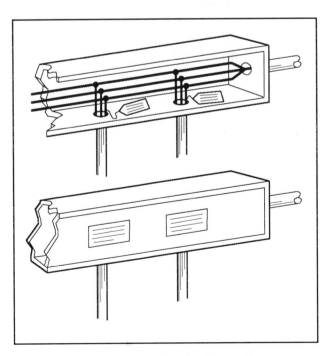

Fig. 10–21 Marking of circuits in auxiliary gutter.

Gutters installed in wet locations must be raintight.

Bare conductors (such as busses) must have a clearance of at least 1 inch from all grounded surfaces, and 2 inches from bare conductors of opposite polarity.

Wire Fills

Gutters are not allowed to contain more than 30 current-carrying conductors (neutral and grounding conductors or control wiring conductors are not counted) at any one cross-sectional area, unless using the correction factors of *Article 310,* Note 8(a) of Notes to the Ampacity Tables.

The total of the cross-sectional areas of all conductors in the gutter may not be greater than 20 percent of the cross-sectional area of the gutter.

Gutters for elevators are not subject to the two preceding requirements.

Chapter Questions

1. Why are surface raceways not allowed to be concealed?

2. Why are wireways that hold sound recording conductors allowed to be concealed?

3. What thickness of metal is required for surface metal raceways that hold the conductors of a 480-volt system?

4. How much of the cross-sectional area of a surface raceway can be used for a splice?

5. What types of surface metal raceways can be installed through floors, walls, or partitions?

6. Can splices be made in underfloor raceways?

7. Can flexible metal conduit be used to connect underfloor raceways to wall outlets?

8. What is the largest size of conductor that can be used in cellular metal floor raceways, without special permission?

9. Is loop wiring considered a splice?

10. Wireways can be installed concealed when containing what type of wiring?

CHAPTER
11

Outlet and Pull Boxes

There are two main articles in the *NEC*®that cover the requirements for boxes: *Articles 370* and *373*. These boxes are important not only because they are important parts of a raceway system, but also because that in them are some of the most important parts of any wiring system—the wiring connections and splices.

The requirements for boxes are as follows.

OUTLET, DEVICE, PULL, AND JUNCTION BOXES, CONDUIT BOXES, AND FITTINGS

(Article 370)

Uses

Round boxes cannot be used to terminate conduits that use locknuts or bushings to connect to boxes.

Nonmetallic boxes can be used only for non-metallic conduits or cables, except when internal bonding means are used for all metal cables or conduits entering the box.

All metal boxes must be grounded.

Only boxes and fittings listed for use in wet locations can be used in these areas. In addition, they must be arranged so that water will not enter or accumulate in the box.

Boxes in hazardous locations must conform to the specific requirements of the article governing the installation. (See *Articles 500* through *517.*)

Unused openings in boxes must be filled. Openings in nonmetallic boxes may be filled with metal fillers, but the fillers must be recessed at least 1/4 inch into the box.

Screws used for attaching a box cannot be used to attach a device used in that box.

No more than a 1/4-inch setback is allowed for boxes installed in noncombustible walls or ceilings. (See Figure 11-1.)

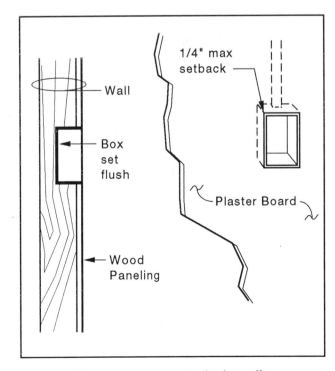

Fig. 11-1 Box setbacks in walls.

Boxes set in combustible walls and ceilings must be set flush with the surface.

Plaster, drywall, or the like must be finished so that no gaps greater than 1/8 inch exist between the box and the plaster, etc.

All boxes must be securely mounted.

Boxes may *not* be mounted on support wires only.

Threaded enclosures without devices that have no more than 100 cubic inches of space can be mounted to two or more conduits, without any other means of support. The conduits must be supported on two or more sides, within 3 feet of the enclosure.

Conduit bodies can be supported by EMT or conduit, as long as their size is not larger than the largest size of conduit or EMT supporting it.

Threaded enclosures containing devices and having no more than 100 cubic inches of space can be mounted to two or more conduits, without any other means of support. The conduits must be supported within 18 inches of the enclosure.

Embedding an enclosure in concrete is considered adequate support.

Nonmetallic boxes can be supported by metal conduits under the same requirements as metal boxes, provided they are specifically listed for this use.

Boxes can be mounted from cord or cable pendants, but must use strain relief connectors (or some other suitable protection).

Every outlet box must have a suitable cover, canopy, or faceplate.

When canopy covers are used, any combustible material under the canopy (such as a wood ceiling) must be covered with noncombustible material (such as plaster). (See Figure 11-2.)

Boxes used for lighting fixture outlets must be designed for that purpose.

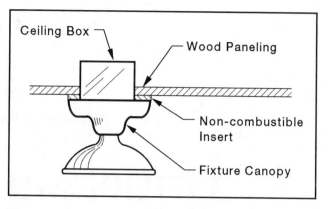

Fig. 11-2 Protection of combustible materials under canopy covers.

Only floor boxes may be mounted in floors. (Raised floors in show windows can use regular boxes if the local authorities permit.)

Boxes alone cannot be used for mounting ceiling fans unless specifically designed for that purpose.

Wire Fills

The amount of wires allowed in an outlet box is a very important consideration. The primary reasons are as follows:

If the box is filled to tightly, the wires will frequently become pinched when a cover is screwed onto the front of the box, causing shorted circuits or damaged conductors.

If the box is too full, the heat of all the conductors could overheat the entire box, raising the temperature beyond the ratings of the conductors.

When splices are crammed too tightly into a box, they can become damaged, interrupting circuits.

The number of wires permitted in a box is calculated based upon the area of the box, measured in cubic inches. (Commercially available boxes are marked, showing their cubic inch area.) The areas required for different sizes of conductors are as follows:

No. 18 AWG	1.5 cu in
No. 16	1.75 cu in
No. 14	2.0 cu in
No. 12	2.25 cu in
No. 10	2.5 cu in
No. 8	3.0 cu in
No. 6	5.0 cu in

Example: If you have 5 No. 12 conductors and 4 No. 10 conductors at an outlet point, the calculations are as follows:

5 No. 12 conductors at 2.25 cu in = 11.25 cu in
4 No. 10 conductors at 2.50 cu in = <u>10.00 cu in</u>
         Total volume required   21.25 cu in

A helpful chart for calculating these box fills is *Table 370-6(a)* of the *NEC*. However, this table does not account for different sized conductors in a box. For such installations (and there are many), you will have to perform calculations such as those shown in the example given above.

The rules for the calculation of box fills are as follows:

One wire must be deducted from the maximum box fill for any fixture stud, cable clamp, or hickey.

Two wires must be deducted from maximum box fill for each mounting strap containing one or more wiring devices (such as receptacles or switches).

When grounding conductors (one or more green wires) are present in a box, one wire must be deducted from the maximum fill. If a second set of grounding conductors (such as an isolated grounding system) is present, one additional wire must be deducted from the maximum fill for this system of grounding conductors also.

A wire that runs straight through a box and is not terminated in the box counts as only one conductor.

The areas (in cubic inches) of plaster rings, raised covers, etc. may be included in the total area of the outlet box that may be filled.

Box Construction

The most commonly used boxes (100 cubic inches or less) are pressed from sheet steel. These are used because they are inexpensive and are sufficient for branch-circuit wiring. Boxes over 100 cubic inches are not necessarily made of thicker metal, but they are subject to different sizing requirements, which keeps the stresses placed upon them within limits. Cast boxes are made of much thicker metal, and are cast rather than pressed into shape. Cast boxes are far stronger than pressed boxes used in more hazardous locations, wet locations, or where threaded boxes are necessary.

Pull Boxes

When conductors No. 4 AWG or larger are contained in cable or conduits 3/4 inch or larger, associated pull or junction boxes must be sized as follows:

1. For straight runs, the length of the pull or junction box must be eight times the trade size of the largest raceway. (If only cables are used [no raceways], the calculations must be made using the sizes of raceways that would be required if the conductors were in raceways rather than in cables.) (See Figure 11-3.)
2. For angle or U pulls, the distance between raceway entries in any wall of the box to the opposite wall must be

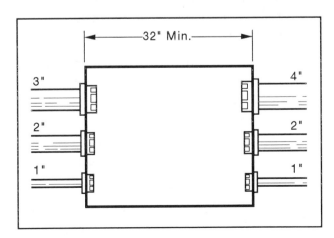

Fig. 11-3 Proper sizing of pull box.

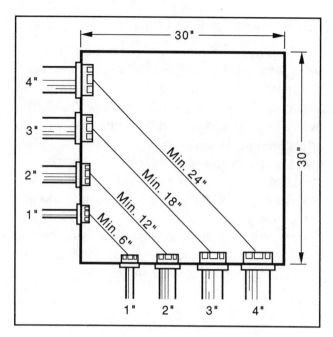

Fig. 11-4 Proper sizing of pull box.

at least six times the trade size of the largest raceway entering that wall. (If only cables are used [no raceways], the calculations must be made using the sizes of raceways that would be required if the conductors were in raceways rather than in cables.) (See Figure 11-4.)

Any pull or junction boxes that have dimensions over 6 feet must have the conductors in the box cabled together or racked. (See Figure 11-5.)

If permanent barriers are installed in a box (as would be the case where communication and power conductors share the same box), each section shall be considered a separate box.

All pull boxes, junction boxes, and conduit bodies must be installed so that the enclosed conductors are accessible following the installation.

Pull or junction boxes for systems of over 600 volts must have one or more removable sides.

Covers for boxes for systems of over 600 volts must be marked "Danger High Voltage Keep Out."

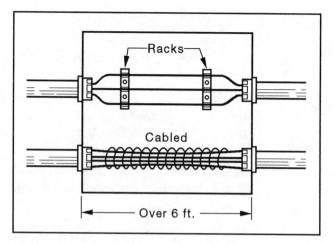

Fig. 11-5 Racking or cabling of conductors in large pull boxes.

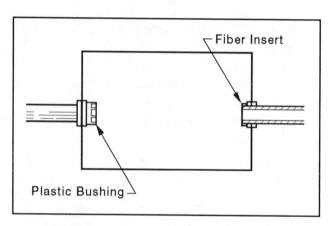

Fig. 11-6 Use of bushings or inserts.

When conductors No. 4 or larger enter a junction box or pull box, insulating bushings, approved fittings, or built-in hubs must be used to protect the conductors. (See Figure 11-6.)

CABINETS AND CUTOUT BOXES

(*Article 373*)

Installation Methods

Cabinets and cutout boxes (including meter fittings) installed in damp or wet locations must be spaced at least 1/4 inch from the surface on which they are mounted. They must also be mounted so that moisture will not enter or accumulate.

No more than a 1/4-inch setback is allowed for boxes installed in noncombustible walls or ceilings.

Boxes set in combustible walls and ceilings must be set flush with, or protruding from, the surface.

Boxes in hazardous locations must conform to the specific requirements of the article governing the installation. (See *Articles 500* through *517.*)

Unused openings in boxes must be filled. Openings in nonmetallic boxes may be filled with metal fillers, but the fillers must be recessed at least 1/4 inch into the box.

When conductors No. 4 or larger enter a cabinet or cutout box, insulating bushings, approved fittings, or built-in hubs must be used to protect the conductors. (See Figure 11-6 above.)

Chapter Questions

1. Can nonmetallic conduits be terminated in nonmetallic boxes?

2. How far must metal knockout fillers be recessed in nonmetallic boxes?

3. Boxes of what size can be supported by only two conduits?

4. May conduit bodies be supported by EMT?

5. How many wires must be deducted from maximum box fills for each device mounting strap?

6. How many wires are deducted from the maximum box fill if two green grounding conductors are present in the box?

7. How many wires are deducted from the maximum box fill if four green grounding conductors are present in the box?

8. For a straight run of 2-inch conduit, what is the smallest allowable pull box that can be used?

9. When are pull boxes required to have removable sides?

10. Are meter fittings considered cutout boxes?

PART
3

POWER
DISTRIBUTION

CHAPTER
12

Feeders

Feeders, covered in *Article 215* of the *NEC®*, are used to distribute fairly large quantities between distribution panel (or service panel) and branch-circuit panels. As with branch circuits, there may be other requirements for specific types of feeders found in other parts of the Code than *Article 215*. In all of these instances, however, all the requirements of *Article 215* apply, unless they are *specifically* excluded.

Feeders are one of the most expensive parts of common wiring systems, requiring large, expensive conduits and large, expensive conductors. The physical routing of feeders is a very important consideration on the job site. Since feeders can easily cost $50 to $150 per foot, the length of a run is a rather serious matter.

The requirements are as follows:

THE SIZES AND RATINGS OF FEEDERS

Most feeders are rated 100 amps or higher. It is possible, however, for feeders to be rated less than this for certain uses.

Feeder conductors must have an ampacity no less than that required to carry their load.

Feeders can be no smaller than 30 amperes when the load being supplied consists of the following types of circuits:

1. Two or more branch circuits supplied by a 2-wire feeder.
2. More than two 2-wire branch circuits supplied by a 3-wire feeder.
3. Two or more 3-wire branch circuits supplied by a 3-wire feeder.
4. Two or more 4-wire branch circuits supplied by a 3-phase, 4-wire feeder.

Feeders must be protected with overcurrent protective devices.

GROUNDING AND OTHER REQUIREMENTS

Since feeders carry large amounts of current, they can supply large fault currents. Because of this, the grounding requirements for them are especially important. Many feeders could supply fault currents of several hundred amps without tripping their circuit breaker. If there were no good grounding system, these very large fault currents could exist (causing great damage and danger) without ever being detected.

Feeders that contain a common neutral are allowed to supply two or three sets of 3-wire feeders, or two sets of 4- or 5-wire feeders.

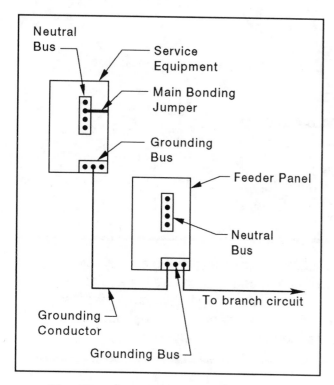

Fig. 12-1 Grounding of feeder panels.

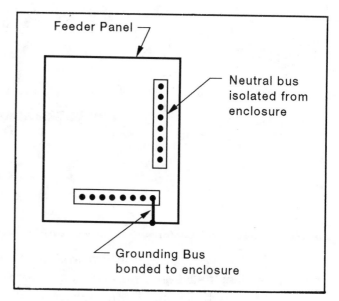

Fig. 12-2 Bonding of grounding bus.

In any metal raceway or enclosure, all feeders with a common neutral must have their conductors run together.

When a feeder supplies branch circuits that require equipment grounding conductors, the feeder must supply a grounding means, to which the equipment grounding conductors (from the branch circuits) can be connected. (See Figures 12-1 and 12-2.)

Two-wire AC or DC circuits of two or more ungrounded conductors are allowed to be tapped from ungrounded conductors of circuits that have a grounded neutral. Switching devices in each tapped circuit must have a pole in each ungrounded conductor.

For a 4-wire delta system, the phase conductor having the higher voltage must be identified by an orange marking.

Feeders that supply 15- or 20-ampere branch circuits that require ground-fault protection can be protected against ground-faults, rather than protecting individual circuits or receptacles.

Chapter Questions

1. Are feeders allowed at under 100 amps?

2. Why is the routing of feeders important?

3. Can a 20-amp feeder be used to supply two 3-wire branch circuits?

4. How many sets of 3-wire feeders can be supplied by a feeder that has a common neutral?

5. How must the high phase of a 4-wire delta system be identified?

6. How are feeders protected from overheating and short-circuit damages?

CHAPTER
13

Services

Article 230 of the *NEC*®covers virtually every aspect of electrical services, the wiring that brings current to a structure or utilization point—the beginning point of a wiring system. Follow through these requirements carefully; it is important that you understand how to handle services.

Service is the part of a wiring system that has the largest capacity, and also the part of the system that generally has the least protection. An average residence may have service conductors rated for 200 amperes, which is all that they are ever likely to carry—but the only overcurrent protective device protecting them is often a pole-mounted fuse rated at several thousand amperes! In other words, the power that flows through most services is for all practical purposes unfused. Remember that this poses an increased danger. Many types of circuits can be shorted and cause a hazard, but in most cases a circuit breaker will trip and eliminate the hazard in a very short time. However, in the case of services, the circuit will provide a practically unlimited amount of current. It will not blow a breaker—it will merely keep flowing until something burns up! So pay attention to the requirements for services; they are very important.

SERVICE CONDUCTORS

Service conductors are not allowed to pass through a building or structure, then supply another building or structure, unless they are encased in 2 or more inches of concrete.

Conductors are not considered to be in a building (although they actually are) in any of the following circumstances:

1. If they are encased in 2 or more inches of concrete.
2. If they are encased in a raceway, then enclosed in 2 inches of brick.
3. If they are in proper transformer vaults.

The only conductors outside of service conductors allowed in service raceways are grounding conductors and load management conductors that have overload protection.

When a service raceway enters from underground, it must be sealed to prevent the entrance of gas. Empty raceways must also be sealed.

Service cables without an overall jacket must be at least 3 feet from windows or similar openings, except that they are allowed with less clearance over windows (rather than next to them).

Service conductors must have enough ampacity to carry the load placed upon them.

Service conductors are not allowed to be smaller than No. 8 copper or No. 6 aluminum. (When services feed only limited loads of single branch circuits, No. 10 copper [equivalent to No. 12 hard-drawn, which the Code actually specifies] can be used.)

The size of the neutral conductor for a service must be at least the following:

1. 1100 kcmil or smaller service conductors—The neutral must be at least as large as the grounding electrode conductor shown in *Table 250-94.*
2. Larger than 1100 kcmil service conductors—The neutral must be at least 12-1/2 percent of the size of the largest phase conductor. If parallel phase, the neutral must be 12-1/2 percent of the equivalent cross-sectional conductor area.

Each service can have only one set of service conductors, except that multiple tenants or occupants of one building can each have their own service-entrance conductors.

One set of service conductors is allowed to supply a group of service-entrance enclosures.

Service-entrance conductors must be of sufficient size to carry their load.

Service conductors cannot be spliced, except as follows:

1. Clamped or bolted connections in meter fittings are allowed.
2. Service conductors can be tapped to supply two to six disconnecting means grouped at a common location.
3. At a proper junction point where the service changes from underground to overhead.
4. A connection is allowed when service conductors are extended from an existing service drop to a new meter, then brought back to connect to the service-entrance conductors of an existing location.
5. Sections of busway are allowed to be connected together to build the service.

SERVICE CLEARANCES

The importance of service clearances is that of safety. The chief dangers are those such as vehicles striking service drops (causing power outages and perhaps electrocutions), persons tampering with the service drops or having accidental contact with tools, machinery, or other objects. The intent of these requirements is to eliminate any such hazards. (See Figure 13-1.)

Service conductors not over 600 volts must have the following clearances over grade:

1. Above finished grade, sidewalks, platforms, etc. from which the conductors could be reached by pedestrians (but not by vehicles), and where the voltage is not more than 150 volts to ground: 10 feet.
2. Over residential driveways and commercial areas not subject to truck traf-

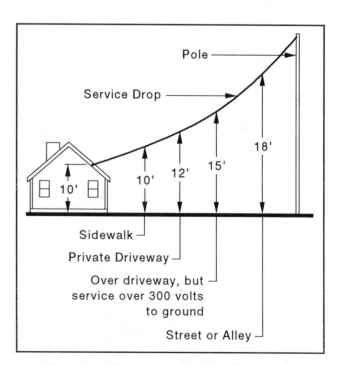

Fig. 13-1 Service conductor clearances.

fic, and where the voltage is not more than 300 volts to ground: 12 feet.

3. For areas in the above classification (12-foot rating) when the voltage is greater than 300 volts to ground: 15 feet.

4. Over public streets, alleys, roads, parking areas subject to truck traffic, driveways on nonresidential property, and other land traversed by vehicles (orchards, grazing, etc.): 18 feet.

5. Conductors not over 600 volts must have an 8-foot clearance over roofs. This clearance must be maintained with 3 feet of the roof surface, measured horizontally.

6. If a roof is subject to pedestrian traffic, it is considered the same as a sidewalk.

7. If a roof has a slope of 4 inches rise for every 12 inches of run or greater, the clearance can be only 3 feet. The voltage cannot be more than 300 volts between conductors. (See Figure 13-2.)

8. If no more than 4 feet of conductors pass over a roof overhang and are terminated by a through-the-roof raceway or approved support, and the voltage is not more than 300 volts between conductors, only 18 inches clearance is required.

9. Horizontal clearance from signs, chimneys, antennas, etc. need only be 3 feet.

10. When these conductors attach to a building, they must be at least 3 feet from windows, fire escapes, etc. (See Figure 13-3.)

The point of attachment of service conductors to a building must be no lower than the above-mentioned clearances, but never less than 10 feet.

Service masts used to support service drops must be of sufficient strength or be supported with braces or guys. (See Figure 13-4.)

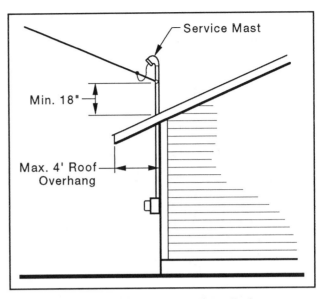

Fig. 13-2 Service mast installation.

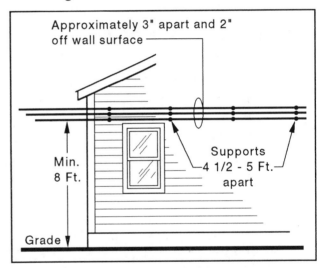

Fig. 13-3 Service conductors run on surface of wall.

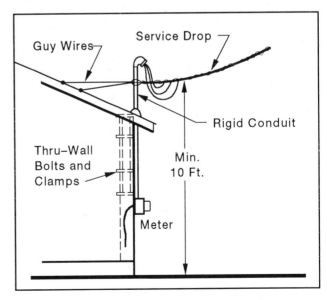

Fig. 13-4 Service mast support.

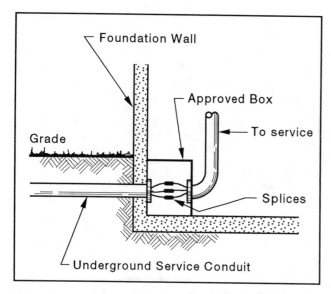

Fig. 13-5 Splicing in underground service run.

UNDERGROUND SERVICE CONDUCTORS

Underground service conductors must be suitable for the existing conditions where they are installed. They must be protected where required. (See Figure 13-5.)

Bare service grounding conductors are allowed as follows:

1. Bare copper conductors in a raceway.
2. Bare copper conductors directly buried, when soil conditions are suitable.
3. Bare copper conductors directly buried, without regard to soil conditions, when installed as part of a cable assembly approved for direct burial.
4. Bare aluminum conductors directly buried, without regard to soil conditions, when installed as part of a cable assembly approved for installation in a raceway or direct burial.

WIRING METHODS FOR SERVICES

Service conductors 600 volts or less can be installed using any of the following methods:

1. Rigid metal conduit.
2. Intermediate metal conduit.
3. Electrical metallic tubing.
4. Service-entrance cables.

5. Wireways.
6. Busways.
7. Cablebus.
8. Open wiring on insulators.
9. Auxiliary gutters.
10. Rigid nonmetallic conduit.
11. Type MC cable.
12. Mineral-insulated, metal-sheathed cable.
13. Liquidtight flexible nonmetallic conduit.
14. Flexible metal conduit; but only for runs of 6 feet or less, between raceways or between raceways and service equipment. An equipment bonding jumper must be run with the conduit.
15. Cable tray systems are allowed to support service conductors.

Service-entrance cables installed near sidewalks, driveways, or similar locations must be protected by one of the following methods:

1. Rigid metal conduit.
2. Intermediate metal conduit.
3. Rigid nonmetallic conduit, when suitable for the location.
4. Electrical metallic tubing.
5. Other approved methods.

Service-entrance cables must be supported within 12 inches of every service head, gooseneck, or connection to a raceway or enclosure. They must be supported at intervals of no greater than 30 inches.

Individual open conductors must be mounted on insulators or insulating supports, and supported as shown in *Table 230-51(c)*.

Cables not allowed to be installed in contact with buildings must be mounted on insulators or insulating supports, and must be supported every 15 feet or less. They must be supported in such a way that the cables will have no less than 2 inches of clearance over the surfaces they go over.

Services must enter exterior walls with an upward slant so that water will tend to flow away from the interior of the building. Drip loops must be made. (See Figure 13-6.)

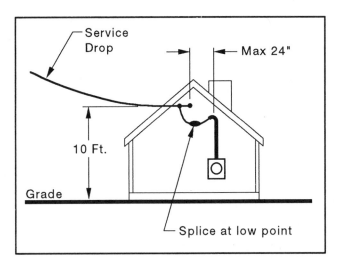

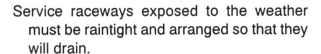

Fig. 13-6 Splices made at low point of loops.

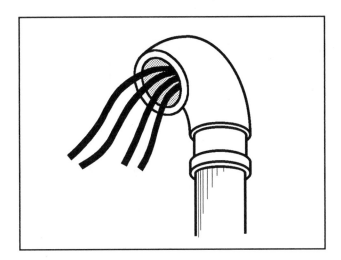

Fig. 13-7 Typical service head.

Service raceways exposed to the weather must be raintight and arranged so that they will drain.

Service raceways must have a raintight service head where they connect to service drops. (See Figure 13-7.)

Service cables, unless they are continuous from a pole to the service equipment, must be provided with a service head or shaped into a "gooseneck." When shaped into a gooseneck, the cable must be taped and painted, or taped with self-sealing, weather-resistant thermoplastic.

Except where it is not practical, service heads for service-entrance cables must be higher than the point of attachment of the service drop conductors.

Drip loops must be made below the level of the service head or the end of the cable sheath.

Service raceways and cables must terminate in boxes or enclosures that enclose all live parts.

For a 4-wire delta system, the phase conductor having the higher voltage must be identified by an orange marking.

SERVICE EQUIPMENT

Energized parts of service equipment must be enclosed and guarded.

A reasonable amount of working space must be provided around all electrical equipment. Generally, the minimum is 3 feet. *Table 110-16(a)* shows specific requirements.

Service equipment must be suitable for the amount of short-circuit current available for the specific installation.

A means must be provided for disconnecting all service-entrance conductors from all other conductors in a building. A terminal bar is sufficient for the neutral conductor. (See Figures 13-8 through 13-10.)

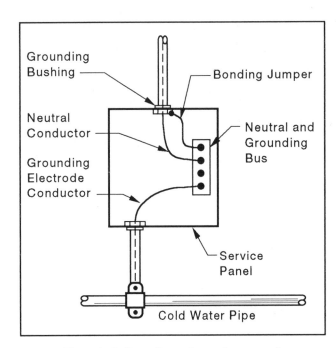

Fig. 13-8 Bonding of service panel.

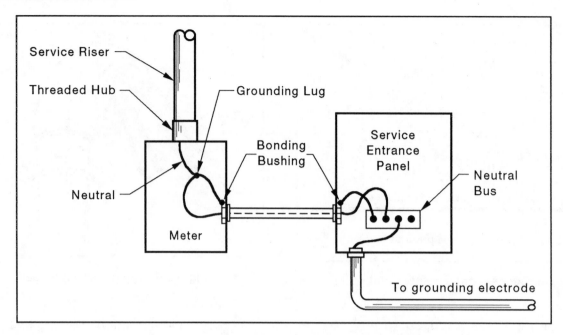

Fig. 13-9 Bonding of meter and service panel.

The service disconnecting means must be installed in an accessible location outside the building, or at the nearest possible point to where the service conductors enter the building, if the disconnecting means is to be installed inside the building.

All service disconnects must be suitable for the prevailing conditions.

The disconnecting means for each service cannot consist of more than six switches or circuit breakers mounted in a single enclosure. (See Figure 13-11.)

Individual circuit breakers controlling multiwire circuits must be linked together with handle ties.

Multiple disconnects must be grouped and marked to indicate the load being served.

Additional service disconnects for emergency power, stand-by systems, fire pumps, etc.

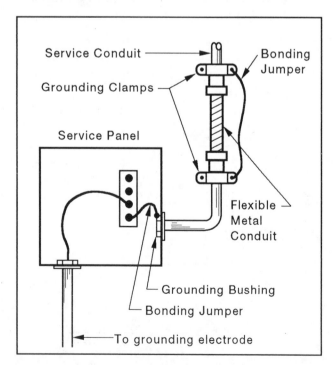

Fig. 13-10 Bonding around flexible link in service conduit.

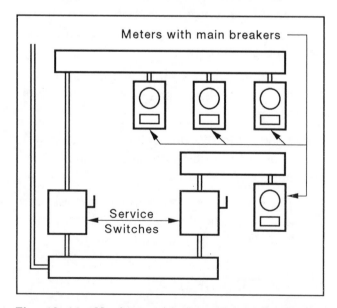

Fig. 13-11 Maximum of six service disconnect switches.

can be separated from other service equipment.

Each occupant in a building must have access to their own service disconnecting means.

A service disconnecting means must simultaneously open all ungrounded conductors.

The service disconnecting means can be operable either manually or by power. When power operable, a manual override for use in the event of a power failure must be made possible.

The disconnecting means must have a rating no less than the load being carried.

Service disconnecting means for one-family dwellings must have a minimum rating of 100 amperes; service disconnects for all other occupancies must have at least a 60-amp rating.

Smaller service sizes are permissible for limited loads, not in occupancies. For loads of two 2-wire circuits, No. 8 copper or No. 6 aluminum conductors can be used. For loads with one 2-wire circuit, No. 12 copper or No. 10 aluminum conductors can be used. These may never be smaller than the branch-circuit conductors.

Only the following items are permitted to be connected to the line side of service disconnects:

1. Cable limiters or current-limiting devices.
2. Meters operating at no more than 600 volts.
3. Disconnecting means mounted in a pedestal and connected in series with the service conductors, located away from the building being supplied.
4. Instrument transformers. (Current—or potential—transformers.)
5. Surge protective devices.
6. High-impedance shunts.
7. Load management devices.
8. Taps that supply load management devices, circuits for emergency systems, fire pump equipment, stand-by power equipment, and fire and sprinkler alarms that are provided with the service equipment.
9. Solar photovoltaic or other interconnected power systems.
10. Control circuits for power operable disconnects. These must be provided with their own overcurrent protective and disconnecting means.
11. Ground-fault protective devices that are part of listed equipment. These must be provided with their own overcurrent protective and disconnecting means.

When more than one building or structure on the same property is under single management, each structure must be provided with its own service disconnecting means. (See Figures 13-12 and 13-13.)

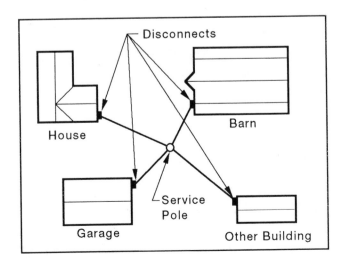

Fig. 13-12 Several buildings fed from one service pole.

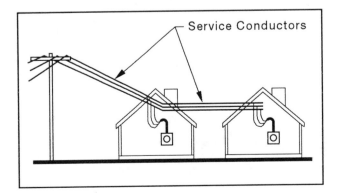

Fig. 13-13 One service drop extended to two buildings.

OVERCURRENT PROTECTION FOR SERVICES

An overcurrent protective device must be installed in each conductor, and must be rated no higher than the ampacity of the service conductor.

No overcurrent device is allowed in a grounded conductor, which would cause a safety hazard.

Services of 100 amps or more for solidly grounded wye systems must have ground-fault protection.

Each occupant in a building must have access to their own service overcurrent protective devices.

Where necessary to prevent tampering, an automatic overcurrent protective device supplying only a single load can be locked.

The overcurrent protective means must protect all equipment except the following:

1. Cable limiters or current-limiting devices.
2. Meters operating at no more than 600 volts.
3. Instrument transformers. (Current — or potential — transformers.)
4. Surge protective devices.
5. High-impedance shunts.
6. Load management devices.
7. Taps that supply load management devices, circuits for emergency systems, fire pump equipment, stand-by power equipment, and fire and sprinkler alarms that are provided with the service equipment.
8. Solar photovoltaic or other interconnected power systems.
9. Control circuits for power operable disconnects. These must be provided with their own overcurrent protective and disconnecting means.

10. Ground-fault protective devices that are part of listed equipment. These must be provided with their own overcurrent protective and disconnecting means.
11. The service disconnecting means.

SERVICES AT GREATER THAN 600 VOLTS

As should be obvious, services that operate at over 600 volts are far more hazardous than services that operate at lower frequencies. Because of this, the requirements for these services are quite a bit more stringent. Safer and heavier methods of wiring are required, and more mechanical protection is required as well. The requirements for these services are shown below. Note that these requirements are in addition to, and in some cases modify the other requirements of *Article 230*.

Service conductors over 600 volts can be installed using any of the following methods:

1. Rigid metal conduit.
2. Intermediate metal conduit.
3. Service-entrance cables.
4. Busways.
5. Cablebus.
6. Open wiring on insulators.
7. Rigid nonmetallic conduit.
8. Cable tray systems that can support service conductors.

Conductors and supports must be of sufficient strength to withstand short circuits.

Open wires must be guarded.

Cable conductors emerging from a metal sheath or raceway must be protected by potheads. (See Figures 13-14 and 13-15.)

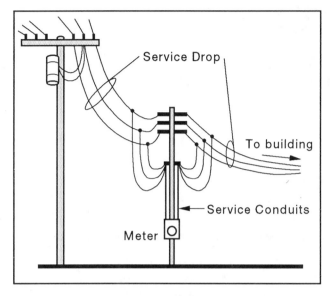

Fig. 13-14 Service with remote meter.

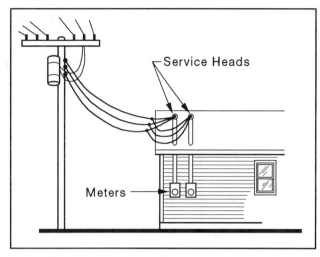

Fig. 13-15 One service drop feeding two meters.

Chapter Questions

1. How close can service conductors be to an apartment window?

2. When are service conductors allowed to pass through buildings?

3. Why must underground service raceways be sealed?

4. Are service-entrance cables allowed to be installed directly above windows?

5. Can a service neutral conductor the same size as its grounding electrode conductor be used?

6. Are clamped connections in meter fittings considered splices?

7. Are bare grounding conductors in raceways allowed?

8. Can 600-volt services be installed in liquidtight flexible nonmetallic conduit?

9. Can services be run through cable trays?

10. How must services be run through exterior walls?

11. What is the maximum number of switches that can be used to disconnect one service?

12. What is the minimum size service for a single-family dwelling?

CHAPTER
14

Busways

Busways (also commonly called "Busduct") are grounded metal enclosures that house insulated (or sometimes, uninsulated) conductors. These conductors are usually in the form of busbars (rectangular copper or aluminum bars), although they can also be in the form of tubes or rods. This entire assembly is assembled in a factory, and only placed and connected in the field.

Busways are used to move fairly large amounts of power (400 to 4,000 amperes) from one place to another. "Plug-in" busways have provisions that allow special disconnect switches to be mounted at almost any point in the run of busway, allowing a great deal of flexibility in tapping power from the busway, as power can be tapped wherever and whenever desired. Taps can also be abandoned whenever desired. Another advantage of busway is that it requires a great deal less labor to install than most alternatives. The most common alternative, individual conductors in a raceway, has a much lower material cost but requires several times as much labor to install.

Busways are covered in *Article 364* of the *NEC®. Sections 364-1, 364-2,* and *364-3* give the scope of the article, a definition of busway, and the requirements for busway installation to comply with general wiring requirements, in addition to the requirements given in *Article 364.*

USES

Section 364-4 specifies where busways may or may not be installed, as follows:

Where Permitted

Where open and visible.

Behind access panels, when:

1. No overcurrent devices are connected to the busway, except for individual devices.
2. Only totally enclosed, nonventilating busway is used.

3. The area behind the access panels is not used for air-handling.
4. The joints between the sections and fittings of the busway are accessible.

Busways may be installed in spaces used for air-handling other than ducts and plenums if they are of the totally enclosed, nonventilating type, having no provisions for plug-in connections.

Where Not Permitted

Where busway might be subjected to corrosive vapors.

Where it might be subjected to physical damage.

In hoistways.

In hazardous locations, except in Class 1, Division 2, and Class 2, Division 2, locations when not enough power is present to ignite specific atmospheres. (See *Section 504-4[b]*, Exception.)

Outdoors or in wet locations, except when identified as suitable for such use.

Installation Methods

The requirements for the installation of busway, detailed in *Sections 364-5, 364-6, 364-7,* and *364-8,* are as follows:

Busways must have supports placed no more than 5 feet apart, except if specifically marked otherwise by the manufacturer.

Unbroken lengths of busway are permitted to pass through dry walls.

Totally enclosed, nonventilating busway may pass through dry floors if it continues to at least 6 feet above the floor.

All open ends of busway must be closed.

When bus enclosures end at machines cooled by a flammable gas, they must use seal-off bushings or baffles so that the gas cannot build up in the enclosure.

Flexible connections or expansion joints must be used in long, straight runs.

All connections and terminations must be accessible after installation.

Branch and feeder circuits from busway may be run by any of the following methods:

1. Busway.
2. Rigid metal conduit.
3. Intermediate metal conduit.
4. Electrical metallic tubing.
5. Rigid nonmetallic conduit.
6. Electrical nonmetallic tubing.
7. Flexible metal conduit.
8. Metal surface raceway.
9. Metal-clad cable.
10. Cord assemblies approved for hard usage. (Allowed only for connection of portable equipment, or equipment being interchanged. Strain reliefs must be installed on the cord.)
11. In instances where the current rating of busway does not exactly match standard ratings of overcurrent devices, the next larger size overcurrent protective device can be used.

All branch-circuit or feeder taps from busway must use plug-ins or connectors that have overcurrent devices in them. This is not required in only the following situations:

1. Where fixtures that have an overcurrent device mounted on them are mounted directly onto the busway.
2. Where an overcurrent device is part of a cord plug for cord-connected fixed or semifixed lighting fixtures.
3. For certain taps. (See *Section 240-21.*) Busway runs over 600 volts must have vapor seals when they enter or leave a building. (This is not required for busways that have air forced through them.)

OVERCURRENT PROTECTION

Sections 364-9, 364-10, 364-11, 364-12, 364-13, and *364-14* detail the overcurrent protection requirements for busways. The pertinent requirements are as follows:

If the overcurrent protective device (fuse or circuit breaker) does not come in a rating that is the same as that of the busway, the next larger size overcurrent protective device can be used.

Busways can be reduced in size in the middle of a run if overcurrent protection is used at the transition point, or if any of the four conditions of *Section 364-11* are met.

When subfeeders or branch circuits are taken from busway runs, the requirements of *Section 364-12* must be complied with.

Busways used as branch circuits must be rated and installed according to the requirements of *Article 210,* which covers branch circuits.

Plug-in busways (which can have circuits tapped from them for their entire length) used as branch circuits must be limited to such a length that the circuit cannot be overloaded.

OVER 600 VOLTS

Busways over 600 volts are covered by Part B of *Article 364 (Sections 364-21* through *364-30).* The installation requirements are as follows:

Busway runs over 600 volts must have fire seals whenever they penetrate a floor, wall, or ceiling.

Drain plugs or drains must be installed at the low point of busways of 600 volts or more to remove excess moisture.

Chapter Questions

1. Can busways be installed behind access panels?

2. What types of busways can be installed through dry walls?

3. Can feeder circuits from busways be installed in flexible metal conduit?

4. Can busways be reduced in size in the middle of a run?

5. When are drains required in busways?

6. Must all busway feeder taps have overcurrent devices?

7. Can busway connections be concealed?

8. What types of joints of connections are necessary for long straight runs?

CHAPTER 15

Switches and Switchboards

The *NEC*®covers the requirements for switches and switchboards in *Articles 380* and *384. Article 380* covers virtually all types of switches, including those used to turn lights on and off, and circuit breakers used as switches. Of course, this article also covers disconnect switches as well. *Article 384* covers switchboards and panelboards. This includes the various types of distribution equipment normally called "switchgear."

SWITCHES
(*Article 380*)

Installation Methods

Switching may *not* be done in grounded conductors.

Switches or circuit breakers cannot disconnect grounded circuit conductors unless all circuit conductors are simultaneously disconnected, or unless the grounded conductor cannot be disconnected until all other conductors are first disconnected. (See Figures 15-1 and 15-2.)

Switches or circuit breakers installed in wet locations must be enclosed in weatherproof enclosures.

Switches (or circuit breakers used as switches) must be mounted so that the center of the operating handle is no more than 6-1/2 feet above the floor or working platform, except in the following cases:

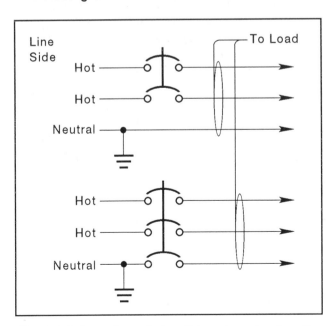

Fig. 15-1 Location of circuit breakers in circuits.

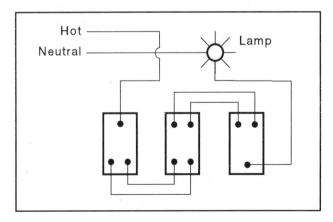

Fig. 15-2 Four-way switching diagram.

1. Fused switches or circuit breakers on busway installations may be mounted at the same level as the busway; however, provision must be made for operating the switch from the floor level.

2. Switches may be installed adjacent to motors, appliances, or other equipment above 6-1/2 feet, but these switches must be accessible by some portable method (such as a stepladder).

3. Isolating switches operated by hooksticks may be mounted higher than 6-1/2 feet.

Switches must be installed so that the voltage between adjacent switches does not exceed 300 volts. If permanent barriers are installed between switches, they are not considered to be adjacent.

Switches mounted in ungrounded metal boxes must have faceplates made of nonconducting and noncombustible material.

Knife Switches

Knife switches are now obsolete, but it is still important to understand their operation and the requirements that pertain to them, since they still exist in some places. If you need more details about these old switches, you can find it in *Section 380-13* of the Code.

The requirements are these:

Single-throw knife switches must be installed so that gravity will not tend to close the switch, or else be equipped with a locking mechanism.

When double-throw knife switches are mounted vertically, they must be equipped with locking mechanisms.

Knife switches must be installed so that the blades are deenergized when the switch is open, except when the load side of the switch is connected to circuits that by their nature may provide a backfeed to the source of power. In these cases, a permanent sign must be placed adjacent to the switch, reading "Warning—Load Side of

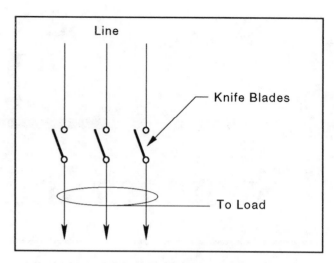

Fig. 15-3 Proper installation of knife switches.

Switch May Be Energized by Backfeed." (See Figure 15-3.)

SWITCHBOARDS AND PANELBOARDS

(*Article 384*)

Installation Methods

Switchboards must have a clear area around them as specified in *Section 110-16*. For switchboards, this area must extend up to the structural ceiling, or 25 feet, whichever is lower. No piping, ducts, or other equipment (except sprinkler systems, which are specifically allowed) may be installed in these spaces. Ventilating equipment serving these areas, necessary control systems, and heavily protected equipment can be installed in these areas.

There must be 3 feet of space between the top of a switchboard and a combustible ceiling, except for totally enclosed switchboards or if a noncombustible shield is installed.

All switchboards must be grounded.

Panelboards installed in wet locations must have raintight enclosures and be arranged so that no moisture will accumulate in the enclosure.

Any panelboards not of the dead front type (which are now obsolete) must be installed in locations where they are accessible only to qualified personnel.

Chapter Questions

1. When can circuit breakers be used as switches?

2. When are barriers between switches required?

3. When are nonconducting switch plates required?

4. How must knife switches be installed?

5. Why are hooksticks sometimes necessary?

6. What is specifically required for backfed knife switches?

7. Can sprinkler systems be installed over switchboards?

8. How much space is required between the top of a switchboard and the ceiling above it?

9. Must all switchboards be grounded?

CHAPTER
16

Transformers

Transformers are devices that transform electrical energy from one circuit to another, usually at different levels of current and voltage but at the same frequency. This is done through electromagnetic induction, in which the circuits never physically touch. The transformer is made of one or more coils of wire wrapped around a laminated iron core. Transformers come in many different sizes and styles.

Transformers are covered in *Article 450* of the *NEC®*. Remember that as far as this article is concerned, the term *transformer* refers to a single transformer, whether it be a single-phase or a polyphase unit.

The requirements of *Article 450* are as follows:

OVERCURRENT PROTECTION

Transformers operating at over 600 volts must have protective devices for both the primary and secondary of the transformer, sized according to *Table 450-3(a)(1)*. If the specified fuse or circuit breaker rating does not correspond to a standard rating, the next larger size can be used.

Transformers operating at over 600 volts that are overseen only by qualified persons can be protected in accordance with *Table 450-3(a)(2)b*. If the specified fuse or circuit breaker rating does not correspond to a standard rating, the next larger size can be used.

Transformers rated 600 volts or less can be protected by an overcurrent protective device on the primary side only, which must be rated at least 125 percent of the transformer's rated primary current. If the specified fuse or circuit breaker rating for transformers with a rated primary current of 9 amperes or more does not correspond to a standard rating, the next larger size can be used. For transformers with rated primary currents of less than 9 amperes, the overcurrent device can be rated up to 167 percent of the primary rated current. (See Figure 16-1.)

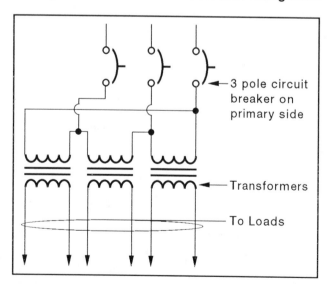

Fig. 16-1 Overcurrent protection on primary side only.

Transformers operating at less than 600 volts are allowed to have overcurrent protection in the secondary only, which must be sized at 125 percent of the rated secondary current, if the feeder overcurrent device is rated at no more than 250 percent of the transformer's rated primary current.

Transformers with thermal overload devices in the primary side do not require additional protection in the primary side unless the feeder overcurrent device is more than six times the primary's rated current (for transformers with 6 percent impedance or less), or four times primary current (for transformers with between 6 and 10 percent impedance). If the specified fuse or circuit breaker rating for transformers with a rated primary current of 9 amperes or more does not correspond to a standard rating, the next larger size can be used. For transformers with rated primary currents of less than 9 amperes, the overcurrent device can be rated up to 167 percent of the primary rated current.

Potential transformers must have primary fuses.

Autotransformers rated 600 volts or less must be protected by an overcurrent protective device in each ungrounded input conductor, which must be rated at least 125 percent of the rated input current. If the specified fuse or circuit breaker rating for transformers with a rated input current of 9 amperes or more does not correspond to a standard rating, the next larger size can

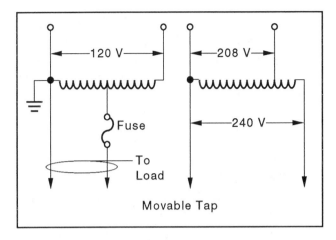

Fig. 16-2 Basic autotransformers.

be used. For transformers with rated input currents of less than 9 amperes, the overcurrent device can be rated up to 167 percent of the rated input current. (See Figures 16-2 and 16-3.)

Installation Methods

Transformers must be installed in places that have enough ventilation to avoid excessive heat build-up.

All exposed noncurrent-carrying parts of transformers must be grounded.

Transformers must be located in accessible locations, except as follows:
1. Dry-type transformers operating at less than 600 volts that are located in the open on walls, columns, and structures do not have to be in accessible locations. (See definition of *accessible* in the glossary of this text.)

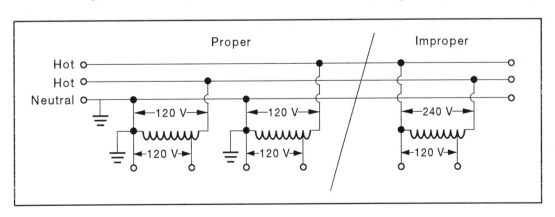

Fig. 16-3 Autotransformer connections.

2. Dry-type transformers operating at less than 600 volts and less than 50 volt-amperes are allowed in fire-resistant hollow spaces of buildings, as long as they have enough ventilation to avoid excessive heating.

Indoor dry-type transformers rated 112-1/2 kVA or less must be separated by at least 12 inches from combustible materials. Fire-resistant, heat-resistant barriers can be substituted for this requirement. Also, such transformers operating at 600 volts or less that are completely enclosed are exempt from this requirement.

Indoor dry-type transformers rated over 112-1/2 kVA must be installed in rooms made of fire-resistant materials. Such transformers with 80°C or higher ratings can be separated from combustible materials by fire-resistant, heat-resistant barriers, or may be separated from combustible materials by at least 6 feet horizontally or 12 feet vertically. Also, transformers with 80°C or higher ratings that are completely enclosed (except for ventilated openings) are exempt from this requirement.

Dry-type transformers installed outdoors must be installed in raintight enclosures, and must not be located within 12 inches of combustible parts of buildings. Transformers with 80°C or higher ratings that are completely enclosed (except for ventilated openings) are exempt from this requirement.

Materials cannot be stored in transformer vaults.

Chapter Questions

1. Are the primary and secondary conductors in a transformer connected?

2. What article of the Code covers transformers?

3. Transformers operating at what voltages are allowed to have overcurrent protection in the secondary only?

4. Where must overcurrent devices be installed for autotransformers rated 250 volts?

5. Why do transformers require ventilation?

6. How far must a 50-kVA dry-type transformer be kept from combustible materials?

7. Can electrical materials be stored in transformer vaults?

8. Do all transformer requirements apply equally to polyphase and single-phase transformers?

CHAPTER
17

Generators

Generators are covered in *Article 445* of the *NEC*®. Remember, however, that a number of other sections of the Code have requirements for generators for specific applications. These sections are frequently those that apply to emergency power applications.

HOW MOTORS AND GENERATORS WORK

Basically, all electric motors and generators operate by using *electromagnetic induction,* which, simply put, is the interaction between conductors, currents, and magnetic fields. Any time an electrical current passes through a conductor (of which copper wires are the most common type), it causes a magnetic field to form around that conductor. This is one of the absolute laws of physics. Conversely, any time a magnetic field moves through a conductor, it *induces* (causes to flow) an electrical current in that conductor. Again, this is an absolute and unchangeable law of physics.

By manipulation of these two laws, in combination with magnetic attraction and repulsion, both generators and motors can be operated. The operation of an electric motor is such that, by intelligent use of electromagnetic induction, electricity turns into physical force, causing the motor to turn. The operation of a generator is such that it turns physical force into an electrical current.

Let us go just one step further and explain in a little more detail how this occurs. The step-by-step operation of an electric motor is thus: An electrical current is turned on and flows through the motor's windings, causing a strong magnetic field to form around the windings. This magnetic field attracts the rotor (the part in the center of the motor that turns—the shaft is at the center of the rotor) and moves it toward the magnetic field, causing the initial movement of the motor. This movement is perpetuated by various means of rotating the magnetic field. Various types of motors do it differently, although the most common method is by using several different windings and sending current to them alternately, thus causing magnetic strength to be in one place one moment, and in another place the next. The rotor will then follow these fields, causing continuous motion.

The step-by-step operation of an electric generator is thus: A stationary magnetic field is set up in a generator's windings by causing a current to flow through them. Then, some type of physical force is used to turn the generator's shaft. On this shaft are mounted coils of wire, which pass through the stationary magnetic field, inducing a current into them. The current is then taken out of the generator (usually by means of a "slip ring" arrangement), and used for some constructive purpose.

While there are any number of variations and modifications to these basic operations, these are the principles by which all motors and generators function. Depending on the type of generator design, we can increase or decrease power, or operate at different voltages. Depending on the type of motor design, we can increase or decrease power, operate at different voltages, or control motor speed.

USE

Generators are most commonly used for very critical or emergency situations. Typical uses would be backup power systems (UPS systems) for important computers, emergency power for hospitals, and emergency power for lighting in theatres and places of assembly.

The requirements for generators are found in *Article 445* of the Code and are as follows:

Locations

Generators must be of a suitable type for the areas in which they are installed.

Overcurrent Protection

Constant voltage generators (which includes virtually all in use) must have overcurrent protection provided by inherent design, circuit breakers, or other means. Alternating-current generator exciters are excepted.

Two-wire DC generators may have only one overcurrent device if it is actuated by the entire current generated, the current in the shunt field excepted. The shunt field is not to be opened.

Generators that put out 65 volts or less and are driven by an electric motor are considered protected if the motor driving them will trip its overcurrent protective device when the generator reaches 150 percent of its full-load rated current.

Installation Methods

The ampacity of conductors from the generator terminals to the first overcurrent protective device must be at least 115 percent of the generator's nameplate rated current. This applies only to phase conductors; neutral conductors can be sized for only the load they will carry. (See *Section 220-22.*)

Live parts operating at more than 50 volts must be protected. Guards are to be provided where necessary.

Bushings must be used where wires pass through enclosure walls.

Chapter Questions

1. What three things are used to make a generator operate?

2. What section of the Code contains the requirements for generators?

3. Do alternating-current generator exciters require overcurrent protection?

4. What is the required ampacity of conductors between generator terminals and the first overcurrent protective device?

5. When are guards required for generators?

6. Generators turn physical force into what?

7. What are typical uses of generators?

PART
4

CIRCUIT WIRING

CHAPTER
18

Branch Circuits

Branch circuits are the last and smallest circuits in the power distribution chain. They are different from other circuits in that they usually feed a number of items, rather than an individual item as feeder and distribution circuits do. Branch circuits are covered by *Article 210,* which applies to branch circuits that supply lighting loads, appliance loads, or combinations of lighting and appliance loads. Requirements for special types of branch circuits are covered in other articles, which apply specifically to those types of equipment. For instance, special branch circuit requirements for motors are given in *Article 430.*

Branch circuits are by far the most common type of wiring installed; therefore, there are a lot of requirements regulating their installation. They are categorized here for ease of use, but you will have to go over them carefully and follow them in your Code book.

CLASSIFICATIONS OF BRANCH CIRCUITS

A branch circuit is rated based upon the setting of its overcurrent protective device. Branch circuits to individual items can have any amperage rating, but branch circuits feeding multiple items must be of one of the following classifications:

1. 15, 20, 30, 40, or 50 amperes.
2. Multioutlet circuits greater than 50 amperes can be used only where they will be accessible only to qualified persons, and not for lighting circuits.

Multiwire circuits can be considered branch circuits as long as all conductors originate from the same panelboard. (See Figures 18-1 through 18-6.)

Multiwire branch circuits are allowed to supply only line to neutral loads unless the

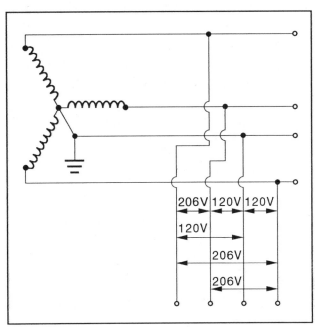

Fig. 18-1 A multiwire circuit.

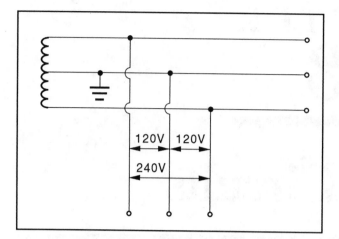

Fig. 18-2 A multiwire circuit.

circuit supplies only one piece of equipment, or if all ungrounded conductors of the circuit open simultaneously by the same branch-circuit overcurrent device.

In dwellings, a multiwire branch circuit that supplies more than one wiring device on the same yoke must have all ungrounded conductors of the circuit open simultaneously by the same branch-circuit overcurrent device.

If more than one voltage system is used in a building, the systems must be separately identified, with the method of identification marked at each panel. (Usually this is done with the colors of the ungrounded conductor's insulation: black, red, blue for 120/208-volt systems; brown, orange, yellow for 277/480-volt systems; etc.)

GROUNDED (NEUTRAL) CONDUCTORS

The grounded circuit conductor (usually called the *neutral* conductor) of branch circuits must be white or natural gray.

If grounded conductors of different systems are installed in common boxes, raceways, etc., the first system must be marked as above, the second system's grounded conductor identified by white insulation with a colored (but *not* green) tracer, and any

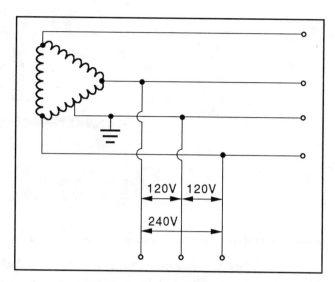

Fig. 18-3 A multiwire circuit.

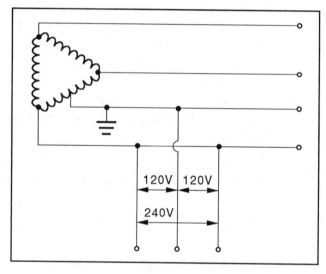

Fig. 18-4 Not a valid multiwire circuit.

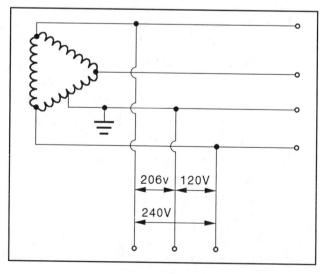

Fig. 18-5 Not a multiwire circuit.

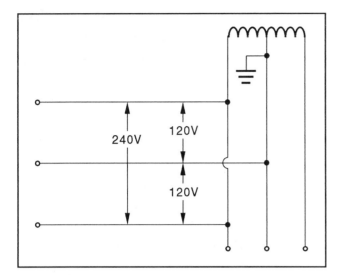

Fig. 18-6 Not a multiwire circuit.

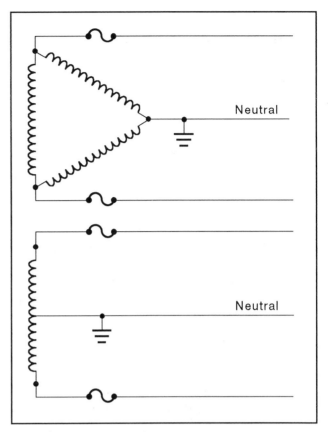

Fig. 18-7 Use of grounded conductors.

other systems identified by their own means.

The equipment grounding conductor of a branch circuit must be either green, or green with a yellow tracer.

The Use of Grounded Conductors

Note that the requirements covering the use of grounded conductors are found in *Article 200,* rather than *Article 210,* as are the other requirements for branch circuits covered in this chapter. These requirements are included with branch circuits because they are applied together, since the vast majority of branch circuits include neutral conductors. (See Figure 18-7.)

The requirements for grounded conductors are these:

All premises wiring systems must have a grounded conductor, except where the *NEC*®specifically permits otherwise.

A grounded conductor must have insulation equal to any ungrounded conductors it is used with.

A grounded premises wiring system must receive its power from a grounded supply system.

A grounded conductor No. 6 or smaller must be covered with white or natural gray insulation, except in the following cases:

1. In fixture wires.
2. Aerial cables can use a ridge on the grounded conductor, rather than a different color of insulation.
3. Where only qualified persons will have access to the conductors, colored conductors can be taped or painted white or gray at their termination.
4. Grounded conductors in Type MI cables can be identified otherwise.

Grounded conductors No. 4 or larger can be identified by either white or gray insulation, or a white marking at their termination.

If grounded conductors of different systems are installed in common boxes, raceways, etc., the first system must be marked as above, the second system's grounded conductor identified by white insulation with a

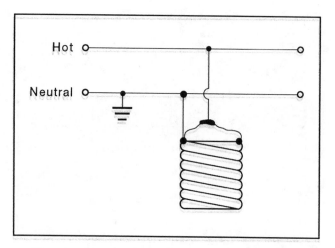

Fig. 18-8 Proper connection of lampholder.

colored (but *not* green) tracer, and any other systems identified by their own means.

Cables to switches can use the grounded (white) conductor to bring power *to* the switch but not *from* it.

Terminals used specifically for grounding conductors must be identified by a sufficiently different color than that of other terminals.

For devices with screw shells, the grounded conductor must be connected to the screw shell, not to the tab. (See Figure 18-8.)

BRANCH CIRCUIT VOLTAGES

Voltages greater than 120 volts between conductors are not allowed in dwellings, hotels, and motels at the following outlet locations:

1. At lampholders.
2. At cord-and-plug connected appliances 1440 watts or less, or less than 1/4 horsepower.

Circuits with voltages between conductors greater than 120 volts are allowed to supply the following:

1. Medium-base screw-shell or other types of lampholders.
2. Electric discharge lighting auxiliary equipment (ballasts).
3. Cord-and-plug or permanently connected utilization equipment.

Circuits between 120 and 277 volts to ground are allowed to supply the following:

1. Electric discharge lighting fixtures that have medium-base screw-shell lampholders.
2. Other types of lampholders that have appropriate voltage ratings.
3. Lighting fixtures that have mogul-base screw-shell lampholders.
4. Electric discharge lighting auxiliary equipment (ballasts).
5. Cord-and-plug or permanently connected utilization equipment.

Circuits between 277 volts to ground and 600 volts between conductors are allowed to supply the following:

1. Electric discharge lighting auxiliary equipment (ballasts) in permanently mounted fixtures. The fixtures must also be at least 22 feet high when mounted on poles, or 18 feet high when mounted on buildings.
2. Cord-and-plug or permanently connected utilization equipment.

RECEPTACLES AND CORD CONNECTORS

Receptacles installed on 15- or 20-amp circuits must be of the grounding type, except to replace receptacles where no grounding means exists.

Grounding contacts of receptacles must be solidly grounded, except where ground-fault interrupter receptacles replace ungrounded receptacles in locations where no grounding means exists, or for vehicle-mounted equipment.

The grounding mentioned above must be done by connection to the equipment grounding conductor.

In locations where more than one type of system (AC or DC, of different voltages, frequencies, etc.) is available, all receptacles must be noninterchangeable, so that a plug for one system will not fit into the receptacle of another system.

GROUND-FAULT CIRCUIT INTERRUPTERS

These devices, commonly called "GFIs," are really modified versions of current transformers used to sense very small fault currents and then interrupt the circuit current. Their effectiveness in protecting life has led to their growing use.

The requirements for GFIs are as follows:

Only ground-fault circuit interrupter-protected 15- and 20-ampere receptacles may be installed in the following locations in dwellings:

1. Bathrooms.
2. Accessible parts of garages.
3. Outdoors, where there is access to the receptacle directly from grade (6 feet from ground or less).
4. Crawl spaces or below grade level in unfinished basements, as receptacles for sump pumps, laundry circuits, or cord-and-plug connected single appliances (usually freezers).
5. As countertop receptacles within 6 feet of a kitchen sink.
6. Boathouses.

All 15- and 20-amp receptacles in hotel and motel bathrooms must have ground-fault protection.

Circuits derived from autotransformers must have ground-fault protection, except for autotransformer output circuits that have a grounded conductor directly connected to the grounded conductor that supplies the autotransformer; or for an autotransformer used to boost an existing 208-volt circuit to 240 volts, and having a grounded output conductor as mentioned above.

Circuits with no grounded conductors are allowed to be tapped from circuits with grounded conductors. Switching devices in the tapped circuits must have a pole in each ungrounded conductor. If multipole switches function as a disconnecting means, all conductors must open simultaneously when the device is activated.

BRANCH CIRCUIT RATINGS

Branch-circuit conductors cannot have an ampacity less than the load connected to them.

Branch-circuit conductors that supply ranges, wall-mounted ovens, cook-tops, or other household cooking appliances must have a rating at least as high as the branch circuit rating and the connected load.

Ranges rated at over 8-3/4 kW must have a branch circuit rated at least 40 amperes.

Tap conductors are allowed from a 50-ampere circuit to a range, wall-mounted oven, cook-top, or other household cooking appliance. They must run no longer than necessary, and must be rated at least 20 amperes.

The neutral conductor of a circuit that supplies a range, wall-mounted oven, cook-top, or other household cooking appliance can be smaller than the ungrounded conductors of the circuit. The size, which must be based on *Table 220-19,* Column A, cannot be smaller than No. 10 copper, and must be rated at least 70 percent of the circuit rating.

Tap conductors are allowed from a 40- or 50-ampere circuit to loads other than household cooking appliances. They must run no longer than necessary, and must be rated at least 20 amperes.

Tap conductors are allowed from a circuit under 40 amperes to loads other than household cooking appliances. They must run no longer than necessary, and must be rated at least 15 amperes.

Branch-circuit conductors and equipment must be protected by overcurrent protective devices.

Heavy-duty lampholders must be used for circuits rated over 20 amperes.

A single receptacle installed on an individual branch circuit must have a rating at least equal to that of the branch circuit.

For branch circuits that supply more than one outlet, the maximum cord-and-plug connected load cannot be more than that shown in *Table 210-21(b)(2)*.

Receptacle ratings for various sizes of circuits are shown in *Table 210-21(b)(3)*. Either 15- or 20-amp rated receptacles are allowed on 20-amp circuits, and only 15-amp receptacles are allowed on 15-amp circuits.

The rating of any cord-and-plug connected appliance cannot be greater than 80 percent of the circuit's rating.

A summary of branch-circuit requirements is shown in *Table 210-24*.

APPLIANCE OUTLETS

Appliance outlets (including laundry outlets) must be installed within 6 feet of the appliance.

Receptacles in dwellings (excluding garages, basements, bathrooms, and hallways) must be installed so that no point along the wall space is more than 6 feet from a receptacle outlet. This includes wall space more than 2 feet in width. Sliding panels in exterior walls need not be considered as wall space. Floor receptacle outlets close to a wall can be considered wall receptacles. (See Figure 18-9.)

In addition to the above requirement, two or more 20-amp receptacle outlets are required for small appliances in a dwelling unit's kitchen, dining room, or breakfast room. (Outlets for dishwashers, disposals, etc. do not count.) No other loads are to be connected to these circuits, except electric clocks or outdoor receptacles.

At least two receptacle circuits are required in every kitchen.

In kitchen and dining areas of dwellings, every counter space 1 foot wide or greater

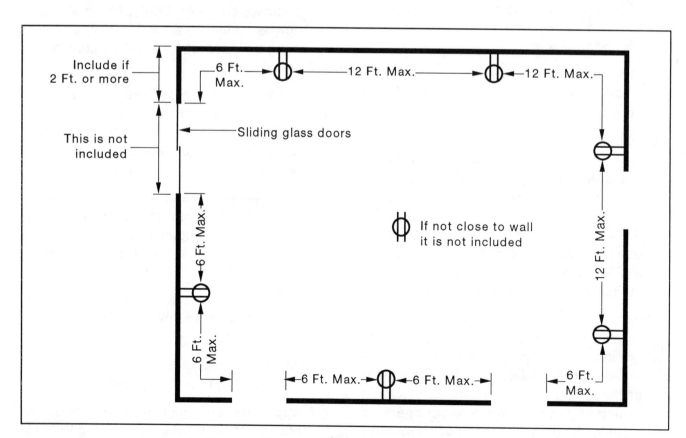

Fig. 18-9 Proper spacing of outlets in rooms.

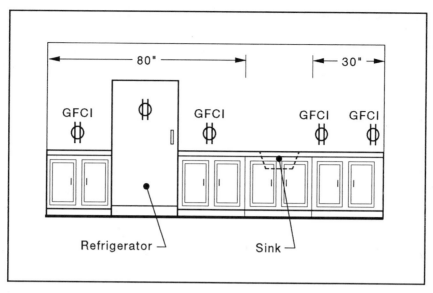

Fig. 18-10 Proper spacing of outlets in kitchens.

must have a receptacle outlet. These receptacles must be placed so that no countertop space is more than 24 inches from a receptacle outlet. All countertop receptacles within 6 feet of the sink must be ground-fault protected. (See Figure 18-10.)

At least one receptacle must be installed in each bathroom, located to the side of the basin. This receptacle must be ground-fault protected.

At least one outdoor receptacle must be installed for all dwelling units. One outdoor receptacle is required for a single-family dwelling unit, two for a two-family dwelling unit. These receptacles must be ground-fault protected.

Dwelling units must have at least one receptacle outlet for a laundry area. (This is not required for apartment buildings that have a common laundry area, or for multifamily dwellings where laundry facilities are not permitted.)

In addition to the above requirement, single-family dwellings must have at least one receptacle outlet in each basement, attached garage, or separate garage with electric power. These receptacles must be ground-fault protected.

Receptacle outlets are required for hallways more than 10 feet long.

Guest rooms in hotels and motels must conform to the receptacle requirements shown above. However, hotels are given more leeway with receptacle spacing requirements to allow for furniture spacing.

Show windows must have receptacles for every 12 inches of window space, measured horizontally.

Except for rooftop equipment for one- and two-family dwellings, receptacles must be installed for the servicing of heating, air-conditioning, and refrigeration equipment. The receptacle must be on the same level as the equipment, within 75 feet. It cannot be connected to the load side of the equipment's disconnecting means.

At least one wall switch-controlled lighting outlet must be installed in each habitable room. Switched wall outlets can be considered lighting outlets for this requirement, except in kitchens.

In addition to the above, switched lighting outlets are required at the entry to an attic, crawl space, utility room, or basement.

Chapter Questions

1. Can black conductors be used as neutrals?

2. What section of the Code says that 277-volt lighting is acceptable?

3. Where are ground-fault interrupters required in existing buildings?

4. Without special permission, what is the largest allowed size of branch circuit serving several loads?

5. How close must a floor outlet be to a wall to be considered a wall receptacle?

6. Can switched receptacles be considered lighting outlets?

7. What is a grounded conductor?

8. What color is used to identify equipment grounding conductors?

9. Why are GFI receptacles used?

10. What is the farthest distance appliance outlets can be from the appliance they feed?

11. Where must a receptacle be installed in a bathroom?

12. How many receptacles are required for show windows?

CHAPTER 19

Outdoor Circuits

The requirements for outdoor branch circuits and feeders are covered in *Article 225* of the *NEC®.* These installations are treated differently than other branch circuits and feeders, due to their differing installation characteristics. Also covered as part of this article is festoon lighting, which is defined as a string of outdoor lights suspended from supports at least 15 feet apart.

Also note that the conductors used for outdoor wiring can carry current at higher levels than conductors installed in cables or conduits. For these installations, you can use the "free air" ratings of *Tables 310-17* and *310-19,* rather than the more commonly used *Tables 310-16* and *310-18.* These tables allow for the use of much larger currents. Beware, however. Conductors sized No. 10 through No. 14 are still required to be protected at their "in conduit or cable" ratings. Refer to the notes of *Tables 310-17* and *310-19,* and you will see this.

OUTSIDE BRANCH CIRCUITS OR FEEDERS

When open individual conductors (for circuits 600 volts or less) are run in overhead spans, the following sizes must be used:
1. For spans 50 feet or less—No. 10 copper, No. 8 aluminum, or larger.
2. Spans over 50 feet—No. 8 copper, No. 6 aluminum, or larger.

Overhead conductors for festoon lighting must be no smaller than No. 12, unless supported by a messenger wire.

Circuits 120 volts between conductors and 277 volts to ground can supply lighting fixtures that illuminate outdoor areas, or commercial or public buildings. The fixtures must be at least 3 feet away from windows, fire escapes, etc.

Wiring 600 volts or under is allowed on buildings, and must be in one of the following:

1. Rigid metal conduit.
2. Intermediate metal conduit.
3. Electrical metallic tubing.
4. Rigid nonmetallic conduit.
5. As open conductors or messenger supported wiring.
6. Multiconductor cables.
7. Type MC or Type MI cables.
8. Cable trays.
9. Cablebus.
10. Wireways.
11. Auxiliary gutters.
12. Flexible metal conduit.
13. Liquidtight flexible metal conduit.
14. Liquidtight flexible nonmetallic conduit.
15. Busways.

Outdoor circuits must enter exterior walls with an upward slant so that water will tend to flow away from the interior of the building.

Open conductors not over 600 volts must have the following clearances over grade:

1. Above finished grade, sidewalks, platforms, etc. from which the conductors could be reached by pedestrians (but not by vehicles), and where the voltage is not more than 150 volts to ground: 10 feet.
2. Over residential driveways and commercial areas not subject to truck traffic, and where the voltage is not more than 300 volts to ground: 12 feet.
3. For areas in the above classification (12-foot rating) when the voltage is greater than 300 volts to ground: 15 feet.
4. Over public streets, alleys, roads, parking areas subject to truck traffic, driveways on nonresidential property, and other land traversed by vehicles (orchards, grazing, etc.): 18 feet.

Conductors not over 600 volts must have an 8-foot clearance over roofs. This clearance must be maintained within 3 feet of the roof surface, measured horizontally.

If a roof is subject to pedestrian traffic, it is considered the same as a sidewalk.

If a roof has a slope of 4 inches of rise for every 12 inches of run or greater, the clearance can be only 3 feet. The voltage cannot be more than 300 volts between conductors.

If no more than 4 feet of conductors pass over a roof overhang and are terminated by a through-the-roof raceway or approved support, and the voltage is not more than 300 volts between conductors, only 18 inches clearance is required.

Horizontal clearance from signs, chimneys, antennas, etc. need only be 3 feet.

When these conductors attach to a building, they must be at least 3 feet from windows, fire escapes, etc.

Buildings three stories or 50 feet in height must have horizontal spaces left between overhead conductor runs 6 feet wide, to allow for fire-fighting ladders.

Live vegetation (trees) cannot be used to support overhead conductor spans, except for temporary wiring.

Chapter Questions

1. What is the required clearance for outdoor circuits over commercial driveways?

2. What is a messenger wire?

3. How long may an outdoor feeder be if it is made with No. 10 copper wire?

4. How must outdoor circuits enter exterior walls?

5. What is the minimum size of aluminum conductor that can be used for 50-foot spans of outdoor circuits?

6. Are cablebuses under 600 volts allowed on the exterior of buildings?

7. Can trees be used to support overhead conductors?

8. When spaces are allowed for fire-fighting access, how wide must these spaces be?

CHAPTER
20

Conductors

The whole of *Article 310* is devoted to conductors, with the intent of making sure that all conductors have sufficient mechanical strength and *ampacity* (current-carrying capacity) and are suited to the environment in which they are installed. Especially critical environmental concerns are temperatures and concentrations of chemicals in the atmosphere.

Make sure that you learn how to use *Tables 310-16* through *310-19* and their associated notes. They are the most important parts of this article, and the parts that you will use far more than any others. The other tables in this article *(Tables 310-61 through 310-84* and their associated notes) are important for certain applications, but you will find that you rarely, if ever, use them. On the other hand, you may use *Tables 310-16* through *310-19* on a daily basis. Again, make sure that you understand every part of these tables and all of the notes that apply to them.

The general requirements of *Article 310* are as follows:

All conductors must be insulated, unless specifically permitted to be bare.

Conductors (copper, aluminum, or copper-clad aluminum) sized No. 1/0 and larger are allowed to be installed *in parallel* (electrically joined together at each end, forming the equivalent of one single conductor). This must be done for all conductors in the circuit—not as one or two conductors in parallel and the others not in parallel.

Conductors in parallel must have the following characteristics:

1. They must be the same length.
2. They must be made of the same material, with the same insulation.
3. The conductors must have the same circular mil area.
4. They must be terminated in the same way.
5. When run in separate raceways, the raceways must have the same characteristics.

Certain specialized systems can also have paralleled conductors. (See *Section 310-4,* exceptions.)

The minimum sizes of wires for different voltages are shown in *Table 310-5.*

Solid dielectric conductors operating at over 2000 volts (for permanent installations only) must be shielded and have ozone-resistant insulation. All metallic shields must be grounded.

Direct burial conductors must be identified for such use. Cables rated over 2000 volts must be shielded. The metal sheath or armor must be grounded.

Conductors in wet locations must be listed for that purpose.

Cables in wet locations must be listed for that purpose.

Cables used for direct burial must be listed for that purpose.

Conductors exposed to oils, vapors, gases, and other corrosive substances must be approved for that purpose.

Conductors cannot be used where their temperature limit can be exceeded.

CONDUCTOR IDENTIFICATION

At one time it was required that all conductors be identified by a color code. In recent years, however, hot conductors have not needed such strict color coding. A number of color coding requirements do still exist, and are extremely important from a safety standpoint. Without such requirements, a lot of electricians would get electrocuted every year.

Please remember the importance of these requirements. *Never ever break them.* To do so would be to endanger the lives of everyone else who works on that wiring system! The old color coding requirements went a long way toward ensuring a high-quality installation and made working on the system far easier. Sometimes engineering companies will write such color coding requirements into specifications for certain projects. When they do, the color code is usually as follows:

120/208-Volt Systems

Phase 1	Black
Phase 2	Red
Phase 3	Blue
Neutral	White
Equipment ground	Green

277/480-Volt Systems

Phase 1	Brown
Phase 2	Orange
Phase 3	Yellow
Neutral	Gray ("natural" or light gray)
Equipment ground	Green with yellow stripe

Following this code is (in many people's eye, including mine) a sign of quality workmanship. You will find that many people (including electrical inspectors) will trust your work far more if you use this coding, even though you are not required by the Code to do so.

The requirements for color coding that *are* in the Code are as follows:

Grounded circuit conductors (neutrals) No. 6 or smaller must be identified by being colored white or natural (light) gray.

An external ridge can be used to identify grounded conductors of multiconductor cables No. 4 or larger.

Bare, covered, or insulated conductors are allowed as equipment grounding conductors.

A conductor No. 6 or larger can be identified as a grounded conductor by a permanent white marking where exposed. This marking can be made with tape, paint, or both.

Ungrounded conductors must be identified by colors other than white, green, or natural gray.

Generally, the ampacities of conductors are found in *Tables 310-16* through *310-19* for common voltages, and *Tables 310-63* through *310-84* for higher voltages. Ampacities can also be calculated (under engineering supervision). Notes 1 through 8 of the Notes to the Ampacity Tables are significant and should be reviewed. *Table 310-16* is most commonly used, giving the ampacities of the most common types of conductors in cables or raceways.

TABLE 310-16

As mentioned at the beginning of this chapter, *Table 310-16* is the most important part of

Article 310. It is the table to which you will refer whenever you must size conductors. Because of this table's importance, I have devoted some space to explaining how to use it.

As you look at this table, you see the various type of conductors (TW, XHHW, THHN, etc.) listed across the top. Also you will notice that they are grouped together according to the ambient temperatures at which they are rated. This is because the amount of heat a conductor's insulation can tolerate determines its ampacity; it can only be rated to run as hot as its insulation can stand.

Beneath each type of conductor are listed the allowable amperages for each size of conductor. Notice also that the left side of the chart covers copper conductors, and the right side covers aluminum.

Now, looking at the left-hand side of the table, note carefully that even though several types of No. 14, No. 12, and No. 10 conductors are rated at more than 15, 20, or 30 amps, respectively, the asterisk next to these conductors indicate that they are not allowed to be used for more than 15 amps for No. 14 wire, 20 amps for No. 12 wire, and 30 amps for No. 10 wire. For aluminum conductors, the same type of restrictions apply, with No. 12 wire rated at a maximum of 15 amps, and No. 10 conductors rated at a maximum of 25 amps.

At the bottom of the table are the correction factors for various ambient temperatures. An example of how these correction factors are used is as follows:

Example: If No. 8 THHN conductors are to be installed in an area with a 50°C ambient temperature, the ampacity of the conductors (as shown in *Table 310-16)* must be multiplied by .82 to account for the 50°C ambient temperature. The calculation would be:

$$55 \times .82 = 45.1 \text{ amps}$$

Therefore, the No. 8 conductor in this case cannot be fused at more than 45 amps.

Table 310-16 applies to installations of not more than three conductors in conduit or cable. If you wish to put more than three conductors in a conduit, or run a cable with more than three conductors (neutrals and grounds do not count in normal circumstances—only hot conductors), the ampacities must be adjusted according to Note 8 to *Table 310-16.* This adjustment is to be done after any necessary ambient temperature adjustments.

While your main concern here is *Table 310-16,* you should know that *Table 310-18* is almost identical to *Table 310-16,* although it gives ampacities for uncommonly used conductors. *Tables 310-17* and *310-19* give ampacities for conductors in *free air* (that is, conductors exposed to air, not enclosed in some type of cable or raceway). Since these conductors are cooled by the surrounding air much better than conductors in conduits or cables, they are allowed much higher ampacities, as you can see from these tables. Be sure not to get mixed up and refer to the wrong tables; there are big differences in ampacities between them.

Chapter Questions

1. What is the ampacity of No. 12 THHN wire in free air?

2. What is the ampacity of No. 4/0 XHHW wire in cable?

3. What is the ampacity of No. 3 THHN wire in free air with an 80°C ambient temperature?

4. Can parallel conductors be of the same size but have different types of insulation?

5. Is color coding required for hot conductors?

6. What is the technical name for hot conductors?

7. At what voltages must conductors be shielded?

8. What type of marking is required for neutral No. 4 or larger?

9. Which tables of the Code give ampacities for free air conductors?

10. What is the usual color code for 277/480-volt systems?

CHAPTER
21

Open Wiring

Open wiring is insulated conductors installed without additional mechanical protection (that is, open to the atmosphere or whatever may be in the area). This type of wiring is not commonly used but is allowed under certain circumstances. While it is generally less expensive to install than other types of wiring, it is also far more vulnerable to mechanical and chemical damage.

One type of open wiring, knob-and-tube wiring, was once used routinely for wiring both houses and factories, and is still often found in older buildings. I suggest that you pay special attention to the requirements for knob-and-tube wiring, as you are likely to run into it at some time and will need to understand the proper methods of working with it.

Pay special attention to the uses permitted for open wiring, as they are limited. Using open wiring inappropriately can be dangerous.

OPEN WIRING ON INSULATORS

This is an exposed wiring method that uses porcelain cleats, knobs, and tubing to protect and support single insulated conductors. This type of wiring is not allowed to be concealed by the structure or finish of a building. It is commonly used for temporary wiring.

The requirements are found in *Article 320* of the *NEC*®, as follows:

Where Permitted

Open wiring is allowed for circuits 600 volts or less for industrial or agricultural establishments, indoors or outdoors, for services, and in places where it could be subjected to corrosive vapors.

Installation Methods

Any of the conductors types in *Article 310* can be used.

The ampacities of open conductors can be found in *Tables 310-17* and *310-19.*

The conductors must be rigidly supported as follows:
1. Within a tap or splice: 6 inches.
2. Within a dead-end connection (receptacle or lampholder): 12 inches.
3. Otherwise: 4-1/2 feet, or closer when necessary.
4. When noncombustible, nonabsorbent spacers are used every 4-1/2 feet or less, No. 8 conductors can be supported every 15 feet. A spacing of 2-1/2 inches between conductors must be maintained.

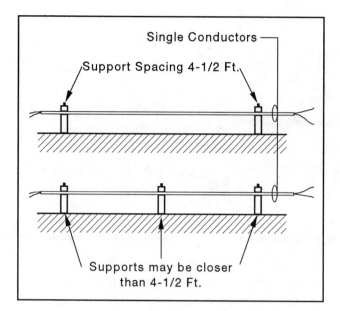

Fig. 21-1 Proper support of open conductors.

5. In mill buildings, if a 6-inch spacing is maintained between conductors, No. 8 and larger conductors can be run across open spaces without support. (See Figure 21-1.)

6. In industrial buildings, conductors No. 250 kcmil or larger can be supported up to 30 feet across open spaces.

Supports must be mounted with at least tenpenny nails, or with screws that penetrate the wood surface to at least one-half the height of the knob.

Wires No. 8 or larger must be tie-wired to their supports with a wire having an insulation equal to that of the conductor.

Conductors can be enclosed in flexible non-metallic tubing in dry locations that are not subjected to damage. The tubing must be in sections at least 15 feet long, and must be supported every 4-1/2 feet.

When conductors penetrate floors or walls, they must be run through an insulating, noncombustible tube (usually made of porcelain). If the tube is too short to reach all the way through the hole, an insulating sleeve can also be used. Each conductor must have its own tube.

Open conductors must be kept at least 2 inches from all other conductors or piping.

If closer, they must be separated by a firmly mounted insulating material. If tubes are used, they must be supported on each end.

Conductors should (this is more of a suggestion than a requirement) pass over, rather than under, other piping systems. (To prevent dripping, etc.)

Drip loops must be made in conductors passing from wet or damp locations into dry locations. The loops should be made on the wet side. When entering, they should go slightly upward, so that water will dry into the wet area rather than the dry area.

Conductors within 7 feet from a floor must be protected by one of the following means:

1. By 1-inch thick guard strips placed next to the wiring.
2. By a running board at least 1/2 inch thick. The sides must be at least 2 inches high.
3. By the above method, with a cover that comes no closer than 1 inch from the conductor.
4. By being enclosed in:
   a. Rigid metal conduit.
   b. Intermediate metal conduit.
   c. Rigid nonmetallic conduit.
   d. Electrical metallic tubing.

   When enclosed in raceways, care should be taken to group the conductors so that the current is approximately equal in both directions.

Conductors in attics or roof spaces accessible via a stairway or permanent ladder must be installed on floor joists, studs, or rafters up to 7 feet from the floor, and protected by running boards that extend at least 1 inch on each side of the conductor.

Conductors in attics or roof spaces not accessible via a stairway or permanent ladder must be installed on floor joists, studs, or rafters. If there is less than 3 feet of headroom in the area, and the wiring is installed before the building is completed, it need not run along the floor joists, studs, or rafters.

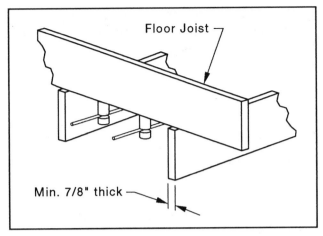

Fig. 21-2 Guarding of open conductors.

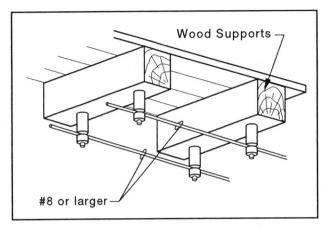

Fig. 21-3 Support of large open conductors.

If surface-mounted snap switches are mounted on insulators, no boxes are required. (See Figures 21-2 through 21-6.)

MESSENGER SUPPORTED WIRING

Messenger supported wiring is a system of exposed wiring that uses a messenger wire (usually a steel cable) to support other conductors. Triplexed and quadruplexed cables are a type of messenger supported wiring.

These systems have been used for many years in industrial installations. They are inexpensive for straight aerial runs. In addition to their general cost-effectiveness, they have the benefit of "free air" ampacity ratings (see *Tables 310-17* and *310-19*), which allows for more current to flow through a given size of conductor. Since the conductors are typically the most ex-

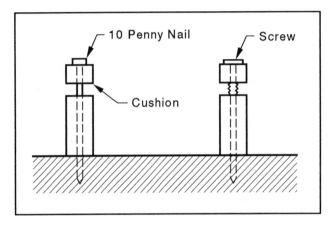

Fig. 21-4 Mounting of supports.

pensive part of a wiring system (in material costs, not the labor required to install the materials), this can amount to a significant savings.

The requirements for messenger supported wiring are given in *Article 321,* as follows:

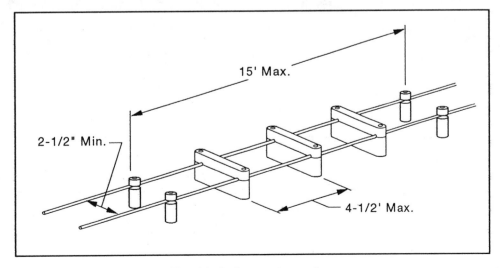

Fig. 21-5 Support spacing.

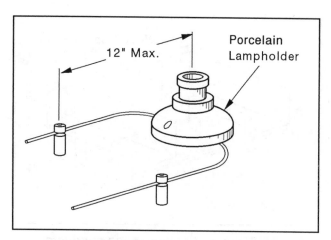

Fig. 21-6 Installation of lampholder.

Where Permitted

The following types of cables can be supported by messenger wires:

1. Mineral-insulated, metal-sheathed cables.
2. Metal-clad cables.
3. Multiconductor underground feeder and branch-circuit cables.
4. Power and control tray cables.
5. Other multiconductor, factory-assembled cables.
6. In industrial establishments only: Any of the conductor types of *Tables 310-13* or *310-62.*
7. In industrial establishments only: Type MV cable.

Messenger supported wiring is allowed in hazardous areas only when specifically permitted. (See *Sections 501-4, 502-4, 503-3,* and *504-20.*)

Where Not Permitted

Messenger supported wiring is not allowed in hoistways, or where it is subjected to physical damage.

Installation Methods

The messenger wire must be supported at ends and at locations in between, to avoid tension on the wiring.

The conductors must be kept from touching the messenger supports, structural members, walls, or pipes.

Messenger wires must be grounded according to the same guidelines as for equipment enclosures.

Any approved and insulated types of splices can be used.

CONCEALED KNOB-AND-TUBE WIRING

This wiring method is similar to open wiring on insulators (*Article 320*), except that it is designed to be concealed within walls and behind other building surfaces.

For many years, this was the method of choice for wiring houses and commercial buildings. Eventually, this method was replaced by Type NM cable, but a great number of knob-and-tube installations still exist. The knobs, tubes, and other materials used in this type of wiring are no longer manufactured; therefore, extensions or repairs are usually made with Type NM cables.

The requirements for knob-and-tube wiring are given in *Article 324* of the *NEC*, as follows:

Where Permitted

In the hollow spaces of walls and ceilings.
In unfinished attics and wall spaces.

Where Not Permitted

In commercial garages.
In theatres and similar locations.
In motion picture studios.
In hazardous locations.
In wall or ceiling spaces where loose, foamed, or foamed-in insulation is present.

Installation Methods

Any of the conductor types of *Article 310* can be used.

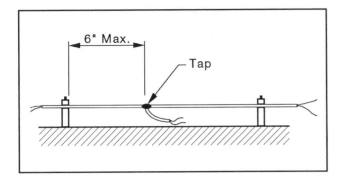

Fig. 21-7 Location of splices or taps.

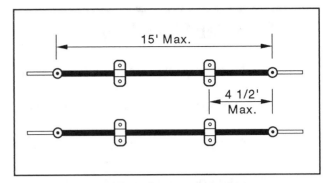

Fig. 21-8 Support spacing.

The conductors must be rigidly supported as follows:

1. Within a tap or splice: 6 inches. (See Figure 21-7.)
2. Otherwise: 4-1/2 feet.

When the conductors must be fished into finished walls, they can be enclosed in unbroken lengths of flexible nonmetallic tubing and do not require support.

Conductors must be tie-wired to the knobs with wires that have insulation equivalent to that of the conductor.

The conductors must be kept at least 3 inches from each other and 1 inch from building surfaces.

In areas (such as at panels) where space is limited, the conductors can be closer to each other than specified above, but must be enclosed in flexible nonmetallic tubing from the area that requires support back to the last support. (See Figure 21-8.)

When conductors pass through floors, walls, or wood cross members, they must be protected by an insulating tube (made of porcelain). The tube must extend through and at least 3 inches beyond the hole.

Open conductors must be kept at least 2 inches from all other conductors or piping. If closer, they must be separated by a firmly mounted insulating material. If tubes are used, they must be supported on each end.

Conductors in attics or roof spaces accessible via a stairway or permanent ladder must be

installed on floor joists, studs, or rafters up to 7 feet from the floor, and protected by running boards that extend at least 1 inch on each side of the conductor.

Conductors in attics or roof spaces not accessible via a stairway or permanent ladder must be installed on floor joists, studs, or rafters. If there is less than 3 feet of headroom in the area, and the wiring is installed before the building is completed, it need not run along the floor joists, studs, or rafters.

Approved types of splices must be used. In-line and strain splices are not allowed. (See Figure 21-9.)

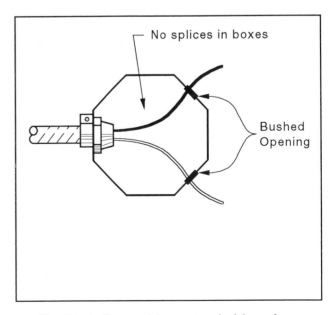

Fig. 21-9 Open wiring extended from box.

Chapter Questions

1. Name one type of messenger supported wiring.

2. Are open wiring systems considered as having free air ratings?

3. Can metal-clad cables be supported by messenger wires?

4. Can control tray cables be supported by messenger wires?

5. Which article of the Code contains the requirements for knob-and-tube wiring?

6. What section of the Code has the requirements for messenger supported wiring?

7. Is open wiring permitted for circuits operating at above 600 volts?

CHAPTER
22

Special Cables

In addition to the commonly used types of cables, such as Types NM, AC, UF, SE, etc., the *NEC*®covers a number of special types of cables used only for certain applications. They are as follows:

FLAT CONDUCTOR CABLES

Flat conductor cables (Type FCC) were specially designed for installation under carpet squares, usually in large office areas. They are used to provide a completely accessible installation (the *NEC*®considers wiring under carpet squares to be accessible). Because such installations are very flexible, their design allows for repeated changes in office layouts.

Before Type FCC cables were developed, expensive and difficult wiring system changes were required every time office layouts were changed. By using Type FCC cables, changing office layouts requires merely pulling up carpet squares, rerouting the cables (rather than installing an entirely new system), and replacing the carpet squares. The advent of modular office dividers made Type FCC cable all the more desirable. (See Figures 22-1 and 22-2.)

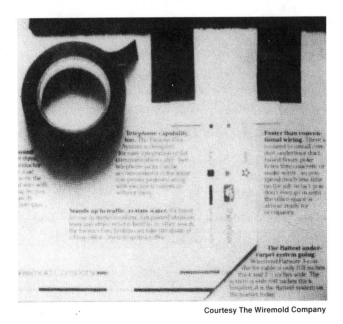

Courtesy The Wiremold Company

Fig. 22-1 Undercarpet wiring system.

Courtesy The Wiremold Company

Fig. 22-2 Undercarpet system receptacle outlet.

The requirements for Type FCC cables are given in *Article 328* of the Code, and are as follows:

Where Permitted

For branch circuits not over 300 volts between conductors or 150 volts to ground.

On finished floors.

On walls, when in surface metal raceways.

In damp locations.

On heated floors, when approved for that purpose.

Where Not Permitted

Outdoors or in wet locations.

Where subjected to corrosive vapors.

In hazardous locations.

In residential, school, and hospital buildings.

Installation Methods

General purpose branch circuits cannot exceed 20 amperes; individual branch circuits cannot exceed 30 amperes.

Floor-mounted Type FCC cable and its fittings must be covered with carpet squares no larger than 36 inches square. The squares must be secured with a release-type of adhesive.

All connectors must be approved types, identified for their use. All bare ends must be insulated.

All floor-mounted Type FCC cables must have both top and bottom shields, over and underneath their entire area.

All metal shields and associated fittings must be electrically continuous to the equipment grounding conductor of the branch circuit.

All receptacles, housings, and devices must be identified for their use, and connected to both the cable and the shields.

Connection to other wiring methods must be made via approved transition fittings.

Type FCC cable must be secured to floors by adhesive or mechanical means.

The crossing of only two cables is permitted.

Type FCC cables can pass over or under a flat communications cable.

Any part of an FCC system that is higher from the floor than 0.09 inches must be tapered.

Cable runs can be added to, provided new cable connectors are used.

Proper polarization of the FCC system must be maintained.

MINERAL-INSULATED, METAL-SHEATHED CABLE

These cables, usually called Type MI cables, are among the best possible wiring methods available. Type MI cables are made of one or more bare copper conductors inside a metal sheath (almost always a copper tube, somewhat similar to the kind plumbers use). These conductors are insulated from the copper sheath and each other by a compressed mineral compound (magnesium oxide powder).

Type MI cables are extremely resistant to heat and chemicals, and are even more resistant to mechanical damage than might first be imagined. They are not, however, recommended for areas that might be subjected to severe mechanical damage, where heavy walled conduits must be used.

While Type MI cables are one of the finest wiring methods available, they are also one of the most expensive. For this reason, they are used primarily in hazardous locations and other areas where less expensive methods of wiring are not allowed or will not function well. (See Figures 22-3 through 22-5.)

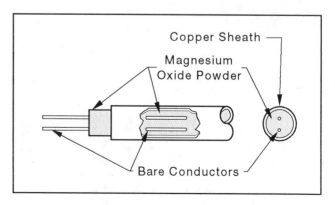

Fig. 22-3 Construction of Type MI cables.

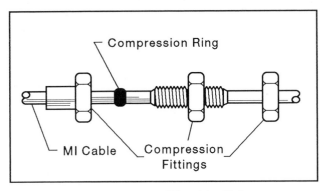

Fig. 22-4 Type MI cable fitting.

The Code covers Type MI cables in *Article 330*. The requirements are as follows:

Where Permitted

For service, feeders, or branch circuits.

For power, lighting, signal, or control circuits.

In dry, wet, or continuously moist locations.

Indoors or outdoors.

Exposed or concealed.

Embedded in masonry.

In any hazardous location.

Exposed to oil or gasoline.

Exposed to corrosive conditions that do not harm its sheath.

Underground, where suitably protected.

Where Not Permitted

In destructively corrosive areas, such as certain chemical facilities.

Installation Methods

In wet locations, and in locations where the walls are frequently washed (dairies, laundries, etc.), mineral-insulated, metal-sheathed cables must be installed with an air space of at least 1/4 inch between them and the surface they are mounted upon.

When Type MI cables are installed through wood structural members, the edges of bored holes must be at least 1-1/4 inches from the edge of the wood framing member.

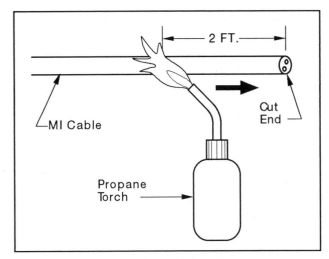

Fig. 22-5 Drying of Type MI cables.

Where the clearance specified above is not possible, a 1/16-inch steel plate must be installed to cover the area of the wiring.

Type MI cables must be supported at least every 6 feet, except when cables must be fished in.

Bends must be made in such a way that the cable is not damaged. The minimum bend radius cannot be less than five times the cable's diameter.

Fittings must be approved for the purpose.

Where single conductors are used, inductive effects must be avoided by one of the following methods:

1. Cutting notches in metal boxes between the holes the conductors pass through.
2. Running the conductors through a box or enclosure wall made of insulating material.

At terminations, seals must be made immediately (to prevent moisture from entering the cable) and the conductors insulated.

SHIELDED NONMETALLIC-SHEATHED CABLES

Shielded nonmetallic-sheathed cables (Type SNM) are essentially nonmetallic sheathed cables (Type NM) that have a second sheath and a layer of shielding made of metal tape and a wire shield. Type SNM cables are used in cable trays and in raceways, and often require special fittings when entering raceways of hazardous areas. They are essentially special-use cables, although they are allowable for a large number of circumstances. But because of their high cost, they are only used for areas they are especially suited to.

The requirements for Type SNM cables are covered in *Article 337,* as follows:

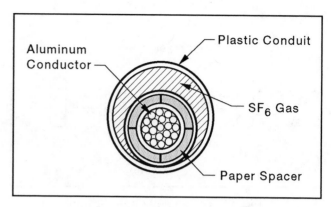

Fig. 22–6 Cross section of Type IGS cable.

Where Permitted

Where the operating temperatures do not exceed the rating of the cables.

In cable trays.

In raceways.

Where permitted, in hazardous locations.

In agricultural buildings, as permitted.

Where Not Permitted

In any areas not specifically listed above.

Installation Methods

Bends must not be so tight as to damage the cable. Five times the diameter is the minimum radius.

Handling of the cable must not damage its covering.

Fittings must be identified for their use.

The wire shield must be bonded to the equipment served and at the power supply point. Bonding can be accomplished by fittings or by other means.

INTEGRATED GAS SPACER CABLE

Integrated gas spacer cables (Type IGS) are factory-made assemblies of insulated conductors enclosed in a flexible nonmetallic conduit that is then filled with a pressurized gas

(sulfur hexafluoride). These cables are used for critical underground cable runs. They are quite expensive, and so are used only in special instances. (See Figure 22–6.)

The requirements for Type IGS cables are found in *Article 325* of the *NEC®,* and are as follows:

Where Permitted

Underground, including direct burial.

For services.

As feeders.

As branch circuits.

Where Not Permitted

For interior wiring.

As interior wiring.

In contact with buildings.

Installation Methods

The radius of any bend must be according to *Table 325-11.*

Individual runs of integrated gas spacer cable cannot have more than four quarter bends (360°).

A valve and cap (for checking and adding gas) must be provided for each length of the cable.

Fittings must be designed for the cable, and must maintain the gas pressure (usually around 20 pounds per square inch).

The ampacity of Type IGS cable is according to *Table 325-14.*

MEDIUM VOLTAGE CABLES

The *NEC*®defines *medium voltages* as being between 2001 and 35,000 volts. (I know it seems that very high voltages are being called "medium," but that is how they are defined in the Code.) These cables (commonly called Type MV cables) are specifically designed for these voltages.

Note that *Article 326,* where these cables are covered, does not cover general wiring requirements for medium voltage work. It merely covers the requirements regarding these cables.

Where Permitted

On power systems up to 35,000 volts.

In wet or dry locations.

In cable trays, only as specifically permitted.

Directly buried.

As messenger supported wiring.

Where Not Permitted

Where exposed to direct sunlight.

The ampacity of Type MV cables must be according to *Tables 310-69* through *310-84.*

POWER AND CONTROL TRAY CABLES

Power and control tray cables (called Type TC cables) are factory-made cable assemblies with insulation characteristics that make them suitable for use in cable trays. The spread of fire is a critical factor for cable tray systems,

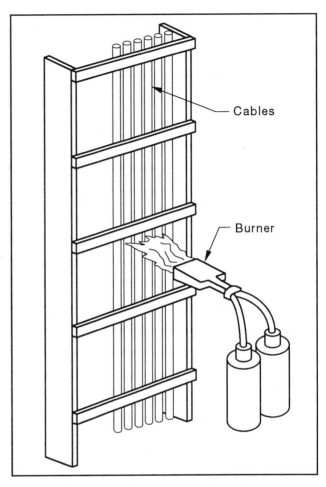

Fig. 22-7 Flame testing of cables.

since they are not enclosed as other wiring methods are. (For instance, if a red-hot rivet falls on a conduit, it will simply bounce off; but if it falls on a cable tray, it will rest on top of the insulated conductors and could easily start a short circuit.) They are, however, frequently used with messenger wires for aerial installations. (See Figure 22-7.)

The requirements for Type TC cables are shown in *Article 340* of the *NEC*®, and are as follows:

Where Permitted

In cable trays.

For power, lighting, or control circuits.

In raceways.

When supported by a messenger wire.

As permitted, in cable trays in hazardous locations.

For Class 1 circuits.

Where Not Permitted

Where subjected to damage.

Installed as open cable on brackets or cleats.

Directly exposed to the sun.

Directly buried, unless specifically identified as suitable.

Installation Methods

Bends must be made in such a way that the insulation is not damaged.

NONMETALLIC EXTENSIONS

Nonmetallic extensions are essentially fairly flat cables intended to provide a cheap method of extending existing branch circuits in areas where other methods would be excessively difficult. They are most commonly used along baseboards, or as an aerial type of wiring supported by messenger cables. (See Figure 22-8.)

The requirements covering the use of nonmetallic extensions are shown in *Article 342* of the *NEC®,* which are as follows:

Where Permitted

From an existing outlet on a 15- or 20-ampere circuit.

Exposed. (See Figure 22-9.)

In dry locations.

In occupied buildings not exceeding three floors above grace.

For aerial cable: In industrial areas that require an extremely flexible wiring system.

Where Not Permitted

As an aerial cable substituting for other aerial wiring methods.

In unfinished attics, basements, or roof spaces.

Where voltage between conductors is over 150 volts for interior use, and 300 volts for aerial cable.

Where subjected to corrosive vapors.

Where run through a floor or partition.

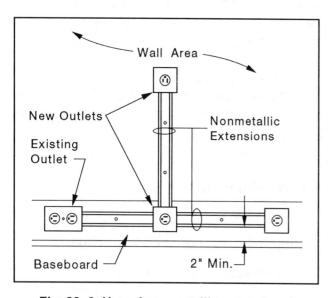

Fig. 22-8 Use of nonmetallic extensions.

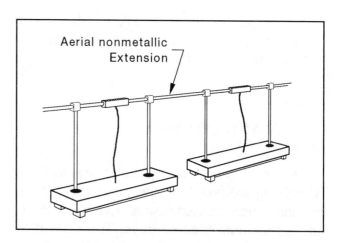

Fig. 22-9 Aerial nonmetallic extensions.

Installation Methods

Splices and taps are allowed only in approved fittings.

Each run must terminate in identified fittings.

Extensions cannot be run within 2 inches of a floor.

Extensions must be secured every 8 inches. (The first attachment can be 12 inches if an attachment plug is used to connect to the outlet.)

No run, no matter how short, can be without support.

Bends that alter the conductor spacing must be covered.

Aerial cables must be supported by messenger cable at each end, and have intermediate supports of 20 feet or less.

Aerial cables must have at least 10 feet of clearance above pedestrian areas, and 14 feet of clearance above areas subject to vehicular traffic.

Aerial cables must have 8 feet of clearance above working surfaces.

Aerial cables can support lighting fixtures when the load is not too much for the supporting messenger cable.

If the messenger cable is properly installed, it can serve as an equipment grounding conductor.

FLAT CABLE ASSEMBLIES

Flat cable assemblies (Type FC cables) are basically three or four conductors formed into a flat cord-like assembly. They are run in U-shaped raceways, and power is tapped from them by the use of special fittings that use a type of knife blade that penetrates through the insulation and into the conductors, taking power from them for small lighting and branch-circuit loads.

These systems are handy in factories and other areas where power may be needed at various points, and where a flexible wiring system is important.

The requirements for flat cable assemblies are given in *Article 363* of the *NEC*®, and are as follows:

Where Permitted

As branch circuits supplying lighting, small appliance, or small power loads.

Exposed.

Where Not Permitted

Where exposed to damage.

Where subjected to corrosive vapors.

In hazardous locations.

In hoistways.

In damp or wet locations.

Installation Methods

Only in identified surface metal raceways ("Unistrut" or "Kindorf").

Splices can be made only in approved junction boxes.

Taps must be made with identified devices.

All dead ends must terminate in an identified end cap.

Only identified fittings can be used.

Support shall be as required for the specific raceways.

Branch circuits cannot be rated more than 30 amperes.

When installed less than 8 feet from the floor, Type FC cables must have a protective cover.

Chapter Questions

1. Can Type MI cable be installed in hazardous locations?

2. Why is Type IGS cable seldom used?

3. What is medium voltage?

4. Can a flat cable assembly be used under floor coverings?

5. What type of mineral is used for Type MI cable?

6. Can Type MI cable be embedded in masonry?

7. Where can Type MI cable not be used?

8. Can Type IGS cable be installed on the surface of a building?

9. Can medium voltage cables operate at 30,000 volts?

10. Can medium voltage cables be installed in cable trays?

11. Can nonmetallic extensions be used for aerial runs?

12. In what type of channel are flat cable assemblies installed?

CHAPTER 23

Cable Trays

Cable trays are assemblies of rigid sections and fittings used to support conductors. The most common types of cable trays are troughs, solid bottom trays, and ladder trays. Cable trays may or may not have covers.

They are used mainly in industrial establishments, and are covered in *Article 318* of the *NEC* ®. The requirements that pertain to cable tray systems are as follows:

Where Permitted

The following types of wiring can be installed in cable trays:

1. Mineral-insulated, metal-sheathed cable.
2. Type MC cable.
3. Power-limited tray cable.
4. Nonmetallic sheathed cable. (Shielded or unshielded.)
5. Multiconductor service-entrance cable.
6. Multiconductor underground feeder and branch-circuit cable.
7. Power and control tray cable.
8. Other approved cables.
9. Any approved conduit or raceway, with the conductors enclosed therein.

The following cables are allowed to be installed in ladder, ventilated trough, and 4- or 6-inch ventilated channel-type cable trays in industrial locations where only qualified persons will have access to the cable trays:

1. Single conductors No. 1/0 AWG and larger. Single conductor cables No. 1/0 through No. 4/0 must be in a venti-

lated cable tray, or a ladder-type cable tray with a maximum rung spacing of 9 inches. Sunlight resistant cables must be used where exposed to the sun.
2. Multiconductor Type MV cables. Sunlight resistant cables must be used where exposed to the sun.

Cable trays can be used as equipment grounding conductors where only qualified persons can service the system.

Cable trays can be used in hazardous locations but must contain only cables approved for such areas.

In corrosive areas that require voltage isolation, nonmetallic cable trays can be used.

Cable trays can be used in environmental air spaces but must contain only cables approved for such areas.

Where Not Permitted

In hoistways.

Where subjected to physical damage.

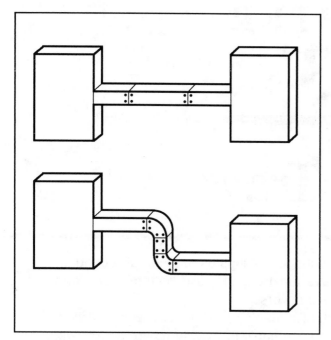

Fig. 23-1 Installation of cable trays.

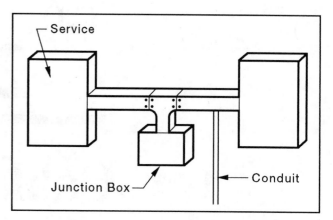

Fig. 23-2 Junction box with cable trays.

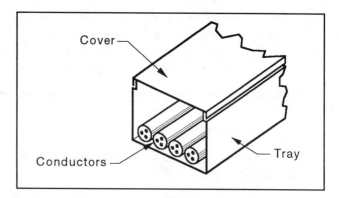

Fig. 23-3 Cross section of cable tray with conductors and cover.

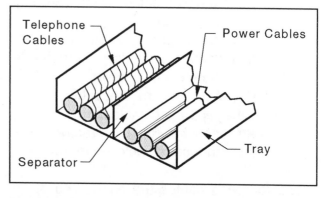

Fig. 23-4 Power and communication conductors in same cable tray.

Installation Methods

Cables rated over 600 volts cannot be placed in cable trays with cables rated 600 volts or under, except for Type MC cables or if divided with barriers.

Cable trays must be installed as a complete system. Any field bends or modifications must not compromise the electrical and mechanical continuity of the system. (See Figures 23-1 and 23-2.)

Complete runs of cable trays must be completed before the installation of any conductors.

Cables must be supported where they enter or leave a cable tray.

Multiconductor cables rated 600 volts or less can be installed in the same cable tray.

Covers or enclosures must be compatible with the cable tray. (See Figures 23-3 and 23-4.)

Cable trays can penetrate walls, floors, partitions, etc., including fire-rated walls and floors, when the openings are sealed with a fire-rated compound. (See Figure 23-5.)

Cable trays must be exposed and accessible. Sufficient space for maintenance must be provided.

Metal cable trays must be grounded.

Steel or aluminum cable trays can be used as equipment grounding conductors. In

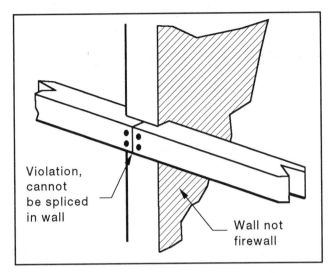

Fig. 23-5 Cable trays cannot be spliced in a wall.

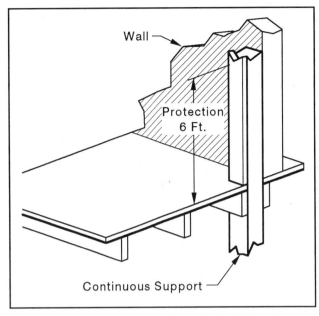

Fig. 23-6 Protection of vertical cable trays.

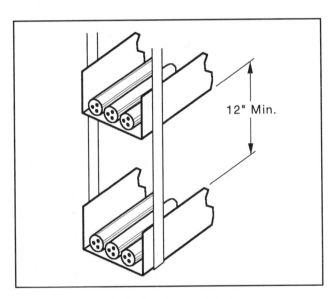

Fig. 23-7 Stacked cable trays.

these cases, the following conditions must be met:

1. The cable tray must be identified for grounding.
2. The cross-sectional area of the cable tray must comply with *Table 318-7(b)(2)*.
3. All sections and fittings must be marked, showing the cross-sectional areas.
4. All sections and fittings must be bonded, using jumpers or mechanical conductors. (See Figures 23-6 through 23-9.)

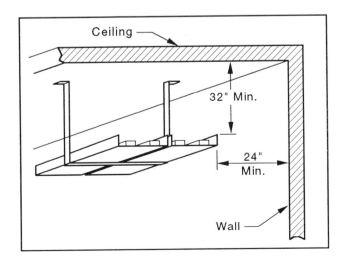

Fig. 23-8 Horizontal and vertical clearances.

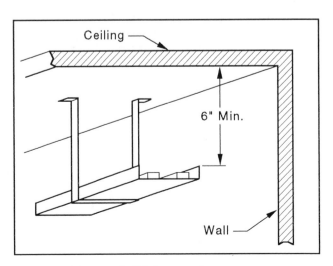

Fig. 23-9 Minimum clearances.

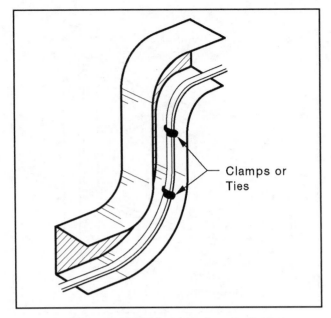

Fig. 23-10 Use of cable ties at elbows.

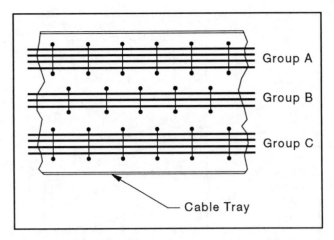

Fig. 23-11 Grouping of conductors in cable trays.

CABLE INSTALLATION

Accessible splices are allowed in cable trays, but they must be made with approved devices.

Cables must be securely fastened to transverse members of the tray. (See Figure 23-10.)

When bushings are installed on raceways feeding conductors into the cable tray, a box is not required.

When parallel conductors are installed, they must be bound together in circuit groups to prevent inductive effects.

Single conductors should also be grouped together. (See Figure 23-11.)

MEMBERS OF CABLES

The amount of multiconductor cables (under 2000 volts) allowed in a cable tray is shown in *Table 318-9* and the following:

1. For ladder or ventilated trays: If all cables are No. 4/0 or larger, the sum of all cable diameters cannot exceed the tray width. The cables must be installed in a single layer.

2. Multiconductor cables can fill up to 50 percent of the cross-sectional area of ladder or ventilated cable trays.

3. For solid bottom trays: If all cables are No. 4/0 or larger, the sum of all cable diameters cannot exceed 90 percent of the tray width. The cables must be installed in a single layer. (See Figure 23-12.)

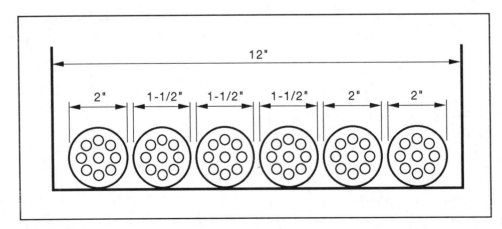

Fig. 23-12 Added multiconductor cable diameters cannot exceed 10.8 inches in a 12-cable tray.

4. Multiconductor cables can fill up to 40 percent of the cross-sectional area of solid bottom cable trays.
5. Multiconductor cables can fill ventilated channel cable trays up to the following:
   a. 3-inch channels: 1.3 square inches.
   b. 4-inch channels: 2.5 square inches.
   c. 6-inch channels: 3.8 square inches.

The amount of single conductor cables (under 2000 volts) allowed in a cable tray is shown in *Table 318-10* and the following:

1. For ladder or ventilated trays: If all cables are 1000 kcmil or larger, the sum of all cable diameters cannot exceed the tray width.
2. For ladder or ventilated trays: If all cables are between No. 1/0 and No. 4/0, the sum of all cable diameters cannot exceed the tray width. The cables must be installed in a single layer.
3. Single conductor cables in ventilated channel cable trays (4-inch or 6-inch): The sum of the cable diameters cannot be greater than the inside width of the channel.

CONDUCTOR AMPACITIES

The ampacities of conductors (under 2000 volts) in cable trays are based on the following:

1. Multiconductor cables (see *Tables 310-16* and *310-18,* Note 8 of the tables): If the cable tray is covered for more than a 6-foot length, the ampacities can be only 95 percent of *Tables 310-16* and *310-18.* If the cables are installed in a single layer in open trays, with at least one cable diameter between cables, the ambient correction factors shown at the bottom of *Table 310-16* can be used.

2. Single conductor cables (including conductors that are triplexed or quadruplexed):
   a. If 600 kcmil or larger: 75 percent of *Tables 310-17* and *310-19,* and Note 8 of the tables.
   b. If between No. 1/0 and 500 kcmil: 65 percent of *Tables 310-17* and *310-19,* and Note 8 of the tables. If the cable tray is covered for more than a 6-foot length, the ampacities can be only 60 percent of *Tables 310-16* and *310-18.*
   c. If No. 1/0 or larger, laid in a single layer in open trays, and with at least one cable diameter between cables, the ampacities of *Tables 310-17* and *310-19* apply.
   d. Where single conductors are installed in a triangular group in open trays, with a spacing of 2.15 conductor widths between circuits, the ampacity of No. 1/0 and larger cables is the same as that for messenger supported conductors.

The ampacities of conductors (over 2000 volts) in cable trays is based on the following:

1. Multiconductor cables (see *Tables 310-75* and *310-76* and Note 8 of the tables): If the cable tray is covered for more than a 6-foot length, the ampacities can be only 95 percent of *Tables 310-75* and *310-76.* If the cables are installed in a single layer in open trays, with at least one cable diameter between cables, ampacities of *Tables 310-71* and *310-72* should be used.
2. Single conductor cables (including conductors that are triplexed or quadruplexed):
   a. If No. 1/0 or larger: 75 percent of *Tables 310-69* and *310-70,* and Note 8 of the tables. If the cable tray is covered for more than a 6-foot length, the ampacities can be

only 70 percent of *Tables 310-69* and *310-70.*

b. If the cables are spaced at least one cable diameter apart, the ampacities of cables No. 1/0 and larger are according to *Tables 310-69* and *310-70.*

c. Where single conductors are installed in a triangular group in open trays, with a spacing of 2.15 conductor widths between circuits, the ampacity of No. 1/0 and larger cables can be up to 105 percent of *Tables 310-71* and *310-72.*

Chapter Questions

1. In what section of the Code are tray cables discussed?

2. Why do certain cable trays need covers?

3. Can Type NM cable be installed in a cable tray?

4. Can cable trays be used as equipment grounding conductors?

5. Are nonmetallic cable trays ever used?

6. Can cable trays be installed in environmental air spaces?

7. For what purpose must space be allowed around cable trays?

8. What methods are commonly used for bonding sections of cable trays?

9. What are cables fastened to in a cable tray?

10. Other than *Tables 310-16* through *310-19,* which tables are used for determining the ampacity of conductors in cable trays?

CHAPTER 24

Receptacle Outlets

The requirements for receptacles, the most common type of wiring device, are found in *Article 410* of the *NEC* ®. They are as follows:

RECEPTACLES

Receptacles used for the connection of portable cords must be rated no less than 15 amps, 125 volts, or 15 amps, 250 volts.

Any receptacles rated 20 amps or less that are to be connected to aluminum wiring must be marked "CO/ALR."

All metal faceplates must be grounded. (This is almost always done automatically, simply by screwing the faceplate onto the grounded yoke of the wiring device.)

Faceplates must seat against the mounting surface (the wall surface) and completely cover any opening.

If the outlet box containing a receptacle is set back from the wall surface, the yoke of the receptacle must be held rigidly against the surface of the wall. (In all cases, the yoke of a receptacle must be rigidly mounted and grounded.)

When receptacles are installed in damp or protected outdoor locations, they must be equipped with a cover that is weatherproof when closed (when a plug is not inserted). Protected locations are considered to be areas under porches, canopies, and the like, where they will not be directly subjected to beating rain or runoff.

Receptacles installed in wet locations must be equipped with a cover that protects the receptacle with the plug inserted, unless a self-closing weatherproof cover is used and only portable tools or equipment are temporarily connected.

Outdoor receptacles are to be located so that water is not likely to touch the outlet cover.

Grounding terminals on receptacles may *never* be used for anything but grounding.

Floor receptacles must be installed so that floor cleaning equipment can be used without damaging the receptacles.

ATTACHMENT PLUGS AND CORD CONNECTORS

Grounding attachment plugs can be used *only* where an equipment ground is present.

Grounding terminals on attachment plugs may *never* be used for anything but grounding.

Chapter Questions

1. What is the lowest amperage rating for receptacles used for connecting portable cords?

2. What does the marking "CO/ALR" mean?

3. Can grounding terminals ever be used for anything other than grounding?

4. When are weatherproof receptacle covers required?

5. Can attachment plugs be used if there is no ground present?

CHAPTER
25

Cable Systems

In almost all circumstances, conductors must be protected. This is accomplished by using some type of protective enclosure—typically a cable, raceway, or conduit—around the conductors.

Conductors in cable are probably the most commonly used type of wiring, being more economical than their alternatives. Cables are, however, somewhat more limited in their use than conduits, because they do not provide as much mechanical protection as conduit or many raceways.

The installation of conductors in cable is very different from that in raceway and has an entirely different set of requirements. In addition, there are many different types of cables—some that are similar to others, and some that are radically different.

One other matter that is important to cover regarding cables is the definition of exactly what a cable is and is not. The *NEC* ®fails to give a general definition of what a cable is (if it did, this definition would be found in *Article 100*), which is perhaps a symptom of the confusion that exists in the field regarding this matter. Typically, we think of a cable as an assembly of one or more conductors enclosed in some type of protective sheath. There are, however, several types of single conductor cables that have no separate sheath. (Single conductor Type UF cables are of this type.) For all practical purposes, such cables are simply single insulated conductors. They may

have a different type of insulation, but they do not appear to have separate conductor insulation and sheath layers.

Because of this lack of definition of terms, differentiating between cables and conductors can be confusing. The best advice I can give is to say that any assembly of conductors that has an outer sheath is certainly a cable. This includes triplexed and quadruplexed conductors, as well as Type UF and similar types of single conductors with special insulations. While these definitions are rather "fuzzy," until the Code committees directly address the subject, this is about as good a definition as anyone can get.

This chapter covers all of the most commonly used types of cables, reserving unusual types for other chapters. The cables discussed in this chapter are the ones that you will most commonly use; therefore, their requirements are the ones that you should know best.

The most important safety considerations are their mechanical protection. Since cables are far more vulnerable than conductors in raceway, this is an important consideration, and the consideration to which the Code gives most of its attention.

ARMORED CABLE

Article 333 of the *NEC*® covers armored cable, which is one of the commonly used types of cables. These cables are used for branch circuit wiring. They have a corrugated metal sheath that is used (along with a metal bonding strip) as the cable's equipment grounding conductor. This use of the sheath requires that all cable connections be very carefully made so that the grounding path is never broken. This is a very serious safety concern, since a break in the grounding system is virtually the same as having no grounding system at all.

Armored cables are also called Type AC cables, or Type ACL cables when the conductors are enclosed in a lead sheath with the corrugated outer sheath over that. The conductors inside of these cables are always insulated, with the exception of the bonding strip, which is not technically a conductor.

The requirements of *Article 333* are the following:

Where Permitted

Type AC cables are allowed as branch circuits and feeders.

Exposed or concealed.

In identified cable trays.

In dry locations.

Embedded in masonry.

Fished in the voids of block walls.

For underplaster extensions.

If used in damp or wet locations, cables with a lead sheath (Type ACL) must be used.

Underground in raceways.

Where Not Permitted

In theatres and similar locations.

In motion picture studios.

In hazardous locations, except as permitted (see *Sections 501-4[b]* and *504-20).*

In areas where corrosive vapors exist.

On cranes or hoists.

In storage battery locations.

In hoistways or elevators.

In commercial garages.

Type ACL cannot be directly buried.

Installation Methods

Type AC cables must be supported within 12 inches of every box, cabinet, or fitting, and every 4-1/2 feet thereafter.

The above requirement is excepted in the following cases:

1. Where the cable is fished into place.
2. Lengths of 2 feet are allowed at terminals where flexibility is required.
3. In lengths of up to 6 feet for connection to light fixtures (or other items) in accessible ceilings.

Bends must not be made so tightly that the cable can be damaged.

Approved insulating bushings must be used at all connections, unless the box or fitting provides equal protection.

Approved connectors must be used for Type AC cables.

When Type AC cables are installed through wood structural members, the edges of bored holes must be at least 1-1/4 inches from the edge of the wood framing member.

Where the clearance specified above is not possible, a 1/16-inch steel plate must be installed to cover the area of the wiring.

Exposed runs of Type AC cables must follow the surface of the building or running board, except as follows:

1. Lengths of up to 2 feet where flexibility is required.
2. On the underside of floor joists in basements. (Must be supported at every joist, and not be subjected to damage.)

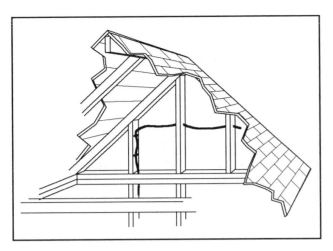

Fig. 25-1 Cable run in attic.

3. In lengths of up to 6 feet for connection to light fixtures (or other items) in accessible ceilings.

Type AC cables in attics or roof spaces accessible via a stairway or permanent ladder, and installed on the top of floor joists, the face of studs, or rafters up to 7 feet from the floor, must be protected by running boards that extend at least 1 inch on each side of the conductor. When attics are not accessible via a stairway or permanent ladder, cables must be protected only within 6 feet of the opening to the attic. (See Figure 25-1.)

When run on the sides of floor joists, studs, or rafters, no running boards are required. (See Figure 25-2.)

METAL-CLAD CABLE

Article 334 covers the requirements of metal-clad cables (Type MC). These cables are often very similar to armored cables. They are defined as "A factory assembly of one or more conductors, each individually insulated and enclosed in a metallic sheath of interlocking tape, or a smooth or corrugated tube." The interlocking tape of Type MC cable is very similar to Type AC cable. The requirements are also similar for the two types of cables; they are as follows:

Where Permitted

For branch circuits and feeders.

For power, lighting, control, or signal circuits.

Directly buried, where identified for such use.

Exposed or concealed.

In cable trays.

As open runs of cable.

As aerial cable on a messenger.

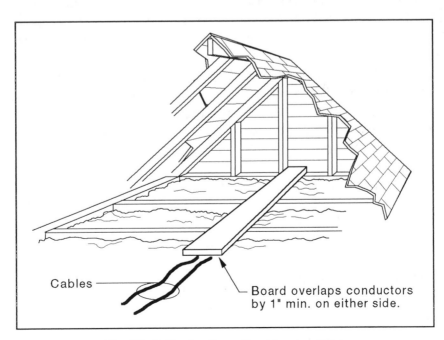

Cables —

— Board overlaps conductors by 1" min. on either side.

Fig. 25-2 Protection of cables in attics.

As specifically permitted for hazardous locations.

In dry locations.

In raceways.

In wet locations in any of the following cases:

1. The metal covering is impervious to moisture.
2. A lead sheath is provided.
3. The conductors are approved for this use.

Where Not Permitted

Where exposed to destructive corrosive vapors.

Installation Methods

Metal-clad cables must be supported at least every 6 feet.

Metal-clad cables in cable trays must comply with *Article 318.*

Bends must be made in such a way that cables will not be damaged.

The minimum radius for various types of cables are as follows:

1. Cables not more than 3/4 inch in diameter: Ten times the cable diameter.
2. Cables between 3/4 and 1-1/2 inches in diameter: Twelve times the cable diameter.
3. Cables greater than 1-1/2 inches in diameter: Fifteen times the conductor diameter.
4. Cables with interlocked armor or corrugated sheaths: Seven times the sheath diameter (external).
5. Cables with shielded conductors: Twelve times the shield diameter, or seven times the cable diameter, whichever is greater.

Fittings must be identified for their use.

Where single conductors are used, inductive effects must be avoided by the following methods:

1. Cutting notches in metal boxes between the holes the conductors pass through.
2. Running the conductors through a box or enclosure wall made of insulating material.

The ampacity of Type MC cables must be according to *Tables 310-16* through *310-19.*

NONMETALLIC-SHEATHED CABLES

Nonmetallic-sheathed cables are the most commonly used cables in the United States and Canada. They are commonly called Type NM cables or "Romex." They are very inexpensive and easy to install, which gives them the lowest installed cost of any widely accepted type of wiring system. These cables have plastic outer sheaths, usually made of some type of PVC. There are also Type NMC cables, which are the same as NM cables but with special outer sheaths, that are acceptable for use around corrosive vapors or chemicals.

Article 336 of the *NEC* ® covers the requirements for Type NM and Type NMC cables. The general safety issue (just as for other types of branch-circuit cables) is mechanical protection. You will find most of these requirements to be similar to those for Type AC cables, with a few more stringent rules, since Type NM cables are not as durable as Type AC cables. (Metal sheaths are obviously stronger than plastic ones.)

The requirements for these cables are as follows:

Where Permitted

One- and two-family dwellings, multifamily dwellings, and other structures.

In identified cable trays.

In dry locations.

In air voids of masonry block walls.

Additionally, Type NMC (corrosion resistant) cable can be used in:

Damp or wet locations.

In masonry block walls.

In a shallow masonry chase, protected with a 1/16-inch steel plate and covered with plaster, etc.

Where Not Permitted

In any structure that exceeds three stories above grade. (The first level, or story, of a building is a level that has half or more of its exterior wall surface at or above finished grade.) If the first floor of a building is used only for parking or storage, it does not have to be counted as one of the three permitted floors. (See Figure 25-3.)

As service-entrance cable.

In commercial garages.

In theatres and similar locations, except where specifically permitted.

In motion picture studios.

In storage battery areas.

In hoistways.

Embedded in masonry.

In hazardous locations, except as specifically permitted.

Type NM cable cannot be installed in the following areas:

Where exposed to corrosive vapors.

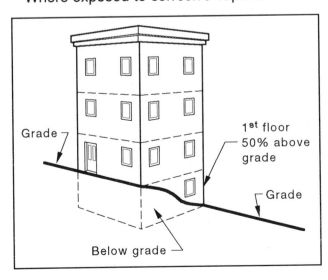

Fig. 25-3 Definition of first floor and basement.

Embedded in masonry.

In a shallow chase in masonry.

Installation Methods

Type NM cables must be supported within 12 inches of every box, cabinet, or fitting, and every 4-1/2 feet thereafter.

The above requirement is excepted where the cable is fished into place.

Devices designed for the termination of Type NM cables without boxes are allowed if the cables are secured within 12 inches of the device. A 6-inch loop of cable must also be left inside the wall.

The cable must follow the surface of the building or running board.

When necessary, the cable must be protected from damage by metal pipe or tubing. When passing through a floor, the cable must be protected at least 6 inches above the floor.

When Type NM cables are installed through wood structural members, the edges of bored holes must be at least 1-1/4 inches from the edge of the wood framing member. Where the clearance specified above is not possible, a 1/16-inch steel plate must be installed to cover the area of the wiring.

Cables with two No. 6 or three No. 8 conductors can be run across joists in unfinished basements without running boards, but they must be secured at every joist. (See Figure 25-4.)

Bends must not damage the cable. The minimum bending radius is five times the cable diameter.

Type NM cables in attics or roof spaces accessible via a stairway or permanent ladder, and installed on the top of floor joists, the face of studs, or rafters up to 7 feet from the floor, must be protected by running boards that extend at least 1 inch on each side of the conductor. When attics are not accessible via a stairway or permanent ladder, cables must be protected only within 6 feet of the opening to the attic.

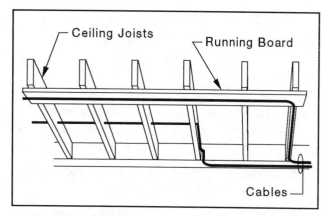

Fig. 25-4 Cables run on basement ceilings.

Switch, outlet, and tap boxes made of insulating material can be used without boxes for exposed wiring or rewiring work where the cables are concealed and fished into place. In these cases, the devices must fit closely around and enclose all stripped parts of the cables. All conductors must terminate; none may be left unattached.

Type NM cables are allowed to be used with nonmetallic boxes.

Type NM cables are allowed to be used with devices with integral enclosures.

SERVICE-ENTRANCE CABLE

Service-entrance cables (also called Type SE cables) are specially designed to be a low-cost method of wiring service entrances. The alternative method, usually individual conductors in raceways, is considerably more expensive. (The additional expense of the raceway-and-conductor method lies in the amount of labor required to install the wiring, not really in the cost of materials.)

These cables are used for services, and also for feeder circuits, most commonly those to ranges and HVAC units. When used this way, some people call them "range cables." Don't let this terminology throw you; "range cables" are nothing more than Type SE cables of appropriate sizes for wiring electric ranges. (See Figures 25-5 through 25-8.)

The requirements for these cables are given in *Article 338,* as follows:

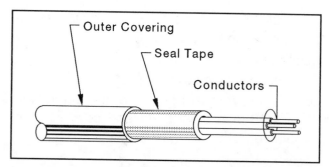

Fig. 25-5 Type SE cable with insulated conductors.

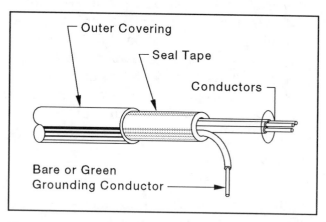

Fig. 25-6 Type SE cable with bare ground conductor.

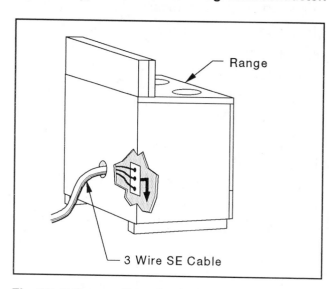

Fig. 25-7 Connection of range with Type SE cable.

Where Permitted

For service entrances, branch circuits, or feeders.

Installation Methods

When used as interior wiring, all of the cable conductors must have rubber or thermoplastic insulation.

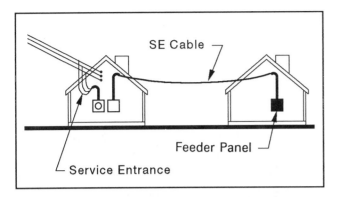

Fig. 25–8 Type SE cable used for feeder.

Type SE cable with an uninsulated neutral cannot be used as interior wiring, unless it has an outer nonmetallic covering and operates at no more than 150 volts to ground. In these cases, it can be used for the following:

1. As a branch circuit that supplies a range, oven, cook-top, or dryer.
2. As a feeder to supply other buildings on the same premises.

Type SE cable cannot be subjected to temperatures higher than the rating of the cable.

Type SE cables must be supported within 12 inches of every box, cabinet, or fitting, and every 4-1/2 feet thereafter.

The above requirement is excepted where the cable is fished into place.

The cable must follow the surface of the building or running board.

When necessary, the cable must be protected from damage by metal pipe or tubing. When passing through a floor, the cable must be protected at least 6 inches above the floor.

When Type SE cables are installed through wood structural members, the edges of bored holes must be at least 1-1/4 inches from the edge of the wood framing member. Where the clearance specified above is not possible, a 1/16-inch steel plate must be installed to cover the area of the wiring.

Cables with two No. 6 or three No. 8 conductors can be run across joists in unfinished basements without running boards, but they must be secured at every joist.

Bends must not damage the cable. The minimum bending radius is five times the cable diameter.

Type SE cables in attics or roof spaces accessible via a stairway or permanent ladder, and installed on the top of floor joists, the face of studs, or rafters up to 7 feet from the floor, must be protected by running boards that extend at least 1 inch on each side of the conductor. When attics are not accessible via a stairway or permanent ladder, cables must be protected only within 6 feet of the opening to the attic.

UNDERGROUND FEEDER AND BRANCH-CIRCUIT CABLE

These cables, usually called Type UF, are among the more commonly used types of cable. They are used for underground runs to devices such as well pumps, post lights, etc. They look very similar to Type NM cables, except that they do not really have a sheath; rather, the conductors are completely enclosed in a special type of plastic resistant to fungus, moisture, corrosion, etc. This extruded plastic covering makes these cables very well suited to burial, but also makes them far more difficult to work with than Type NM cables. They are especially difficult to strip.

The requirements for these cables are given in *Article 339,* and are as follows:

Where Permitted

Underground, including direct burial.

As feeder or branch-circuit cable.

In wet or dry locations.

In corrosive locations.

Where Not Permitted

In theatres.

In commercial garages.

As service-entrance cables.

In motion picture studios.

In storage battery areas.

In hoistways.

In hazardous locations.

Embedded in masonry.

Where directly exposed to sunlight.

Installation Methods

If installed as single conductor cables, the conductors of circuits must be placed close to each other in a trench.

Only multiconductor Type UF cables can be used in cable trays.

Type UF cable cannot be subjected to temperatures higher than the rating of the cable.

Type UF cables must be supported within 12 inches of every box, cabinet, or fitting, and every 4-1/2 feet thereafter.

The above requirement is excepted where the cable is fished into place.

The cable must follow the surface of the building or running board.

When necessary, the cable must be protected from damage by metal pipe or tubing. When passing through a floor, the cable must be protected at least 6 inches above the floor.

When Type UF cables are installed through wood structural members, the edges of bored holes must be at least 1-1/4 inches from the edge of the wood framing member. Where the clearance specified above is not possible, a 1/16-inch steel plate must be installed to cover the area of the wiring.

Cables with two No. 6 or three No. 8 conductors can be run across joists in unfinished basements without running boards, but they must be secured at every joist.

Bends must not damage the cable. The minimum bending radius is five times the cable diameter.

Type UF cables in attics or roof spaces accessible via a stairway or permanent ladder, and installed on the top of floor joists, the face of studs, or rafters up to 7 feet from the floor, must be protected by running boards that extend at least 1 inch on each side of the conductor. When attics are not accessible with a stairway or permanent ladder, cables must be protected only within 6 feet of the opening of the attic.

Chapter Questions

1. Why are running boards sometimes required?

2. What are the differences between Type AC and Type MC cables?

3. What is the difference between a basement and a first floor?

4. What is the difference between Type AC and Type ACL cables?

5. Can Type AC cable be used in motion picture studios?

6. How close to a box must Type AC cable be supported?

7. What article of the Code covers metal-clad cable?

8. What are the most common types of cables used in interior wiring?

9. Can Type NM cable be installed in masonry?

10. Can Type SE cable be used for feeders?

11. What is the difficulty with using Type UF cable?

12. Can Type UF cable be used around storage batteries?

PART
5

EQUIPMENT

CHAPTER 26

Lighting Fixtures

Lighting fixtures are the most common devices connected to electrical circuits. The requirements for the manufacture of lighting fixtures are very strict, and as a result, it is very seldom that a wireman encounters any confusion as to how a light fixture is to be installed. There is some confusion regarding specific applications, however. Specifically, lighting fixtures in hazardous areas or clothes closets are commonly misunderstood.

The *NEC*®'s requirements for lighting fixtures are as follows:

LIGHTING FIXTURES, LAMPHOLDERS, AND LAMPS

(*Article 410*)

Installation Methods

Any fixture weighing more than 50 pounds must be supported independently of an outlet box.

Fixtures must be installed so that the wiring to the fixture can be inspected without being disconnected. (Fixtures connected with flexible cords are excepted.)

Fixtures can be used as raceways only in the following cases:
1. If the fixtures are marked as suitable for this use.
2. Fixtures that connect end-to-end, forming a continuous raceway (or fixtures connected by recognized wiring methods) can carry a 2-wire or multiwire circuit through the fixtures. One additional 2-wire circuit may also be installed if it feeds fixtures that are also connected to the previously mentioned 2-wire or multiwire circuit. (In such cases, these fixtures will have two different sources of power.)

When wires are installed within 3 inches of a ballast, they must have at least a 90°C rating (for example, Types THHN or XHHW.)

Parts of suspended ceilings used to support lighting fixtures must be secured to each other and to the structural ceiling.

Recessed fixtures must have at least a 3-inch clearance from thermal insulation and a 1/2-inch clearance from combustible material, unless specifically listed for such use. Fixtures listed for this use are usually listed as T-type fixtures, the *T* signifying that they are suitable for use in contact with thermal insulation.

Remote fixture ballasts cannot be installed in contact with combustible materials.

Whenever raceway fittings are used to support fixtures, they must be able to support the entire weight of the fixture and any associated lamps, shades, etc.

Fixtures installed in wet locations must be arranged so that water will not enter into or accumulate in them. Fixtures installed in wet or damp locations must be listed and marked as suitable for these locations.

Fixture installations in concrete or masonry that is in contact with the earth must be considered wet locations.

Fixture installations in areas such as basements, cold storage warehouses, and some barns must be considered damp locations.

Fixtures installed in corrosive locations (where chemicals or corrosive vapors are present, such as near gasoline dispensing equipment) must be suitable for the purpose. (If they are not constructed of materials that can withstand corrosive factors, they could quickly fail and cause short circuits or other hazards. Protection is usually provided by enclosing the fixture in a glass bulb.)

When fixtures are to be installed in cooking hoods in nonresidential locations, the following requirements must be met:

1. The fixtures must be vaportight, and the diffusers must be thermal-shock resistant.
2. The fixtures must be identified as suitable for the purpose, and the materials being used must be suitable for the heat encountered in the area where it is installed.
3. The exposed parts of the fixture within the hood must be corrosion resistant, and must have smooth edges so that grease will not accumulate.
4. The wiring supplying the hood (including cable or raceway) must not be exposed within the hood.

No pendant, hanging, or cord-hung fixtures can be installed within 3 feet horizontally, or 8 feet above, any bathtub rim.

All fixtures must be installed so that no combustible materials will be subject to temperatures over 90°C (194°F).

When lampholders are to be installed above highly combustible materials, they must not be of the switched type (with pull-chain), and if individual switches are not installed for each lampholder, they must be at least 8 feet above the floor and protected or located so that the lamps (bulbs) cannot be easily removed or damaged. (See Figures 26-1 and 26-2.)

In show windows, only chain-hung fixtures are allowed to be externally wired.

When fixtures are to be installed in coves, the coves must have enough space to properly install and maintain the fixtures.

Branch circuits are not permitted to pass through incandescent fixtures unless the fixtures are marked as suitable for that use.

All outlet boxes used for lighting must have a suitable canopy or cover.

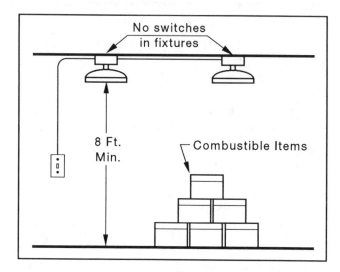

Fig. 26-1 Fixtures above combustible materials.

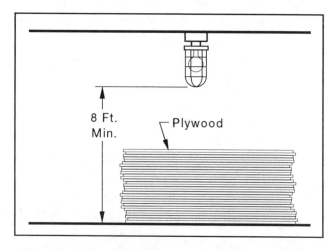

Fig. 26-2 Guarded fixtures over combustibles.

Any combustible wall surfaces (such as wood ceilings) under fixture canopies must be covered with a noncombustible material (such as plaster).

When electric-discharge lighting fixtures (fluorescent, sodium, mercury, or metal halide) are supported independently of outlet boxes, they can be connected with any of the following:

1. Metal raceway.
2. Nonmetallic raceway.
3. Type MC cable.
4. Type AC cable.
5. Type MI cable.
6. Nonmetallic-sheathed cable.
7. Flexible cord (but only if the cord is entirely visible, not subjected to strain or damage, and terminated in a grounded attachment plug or busway plug).

All fixtures (except those having no exposed conductive parts) must be grounded.

Whenever a grounded circuit conductor (neutral wire) connects to a screw-shell lampholder, it must connect to the screw shell rather than the tab inside the lampholder.

Only necessary splices and connections are allowed to be made within fixtures.

No splices or taps can be made in fixture stems, arms, or similar locations.

Up to six cord-connected showcases can be connected to a permanent receptacle outlet. (The showcases are to be connected together with cord and locking connectors.) The cord conductors must be at least as large as the branch-circuit conductors, a grounding type receptacle must be used, and no equipment outside of the showcases must be connected to the cord-supplied circuits. Also, the cases must be no more than 2 inches apart, the first showcase must be within 12 inches from the receptacle, the wiring cannot be subject to damage, and if a free lead is installed at the end of a run, it must be a female cord end and cannot extend beyond the end of the case.

For paired fixtures with one ballast supplying lamps in both fixtures, the wiring from a ballast to a remote fixture can be run up to 25 feet in 3/8-inch flexible metal conduit. Fixture wire carrying line voltage that supplies only the ballast in one of the paired fixtures may also be installed in this raceway.

FIXTURES IN CLOTHES CLOSETS

Installations of various types of lighting fixtures in clothes closets are allowed according to how close they are to what is defined as *storage space.* Storage space is defined as the area within 24 inches of the back and side walls of the closet, and from the floor up to the highest clothes hanging rod, or 6 feet (whichever is higher); the area within 12 inches of the back or side walls above the highest rod, or 6 feet (whichever is higher); and, if the shelves in this area are wider than 12 inches, the area above the shelves (whatever their width).

The following types of fixtures are permitted in closets:

Surface-mounted or recessed incandescent fixtures that have a completely enclosed bulb.

Surface-mounted or recessed fluorescent fixtures.

Fixture Types Not Permitted

Incandescent fixtures with open or partially exposed lamps may not be used in closets.

Pendant fixtures and lampholders (pull-chain or keyless fixtures) are also prohibited.

Locations

Fixtures in clothes closets must be installed as follows:

Surface-mounted incandescent fixtures can be mounted on the wall above a door as long as there is a distance of at least 12 inches between the bulb and the storage area.

Surface-mounted fluorescent fixtures can be mounted on the wall over a door as long as there is a distance of at least 6 inches between the bulb and the storage area.

Recessed incandescent fixtures that completely enclose the lamp (the Fresnel-type may be used, but the baffle-type cannot) can be installed in the wall or ceiling as long as there is a distance of at least 6 inches between the surface of the fixture and the storage area.

Recessed fluorescent fixtures can be mounted in the wall or ceiling as long as there is a distance of at least 6 inches between the surface of the fixture and the storage area.

TRACK LIGHTING

Lighting track must be permanently attached and connected to a branch circuit.

The branch circuit that supplies a lighting track must have a rating equal to or greater than that of the lighting track.

The connected load on a lighting track cannot be greater than the rating of the branch circuit that feeds the track. (This can easily happen if too many track fixtures, or heads, are connected to the track. In reality, there is no way for the installer to ensure that someone will not install too many heads on the track once he has gone. Because of this, it is important to make sure that lighting tracks have plenty of capacity.)

Track lighting cannot be installed in the following locations:

1. In damp or wet locations.
2. In storage battery rooms.
3. In corrosive areas.
4. In hazardous locations.
5. Where concealed.
6. Through walls or partitions.
7. Within 5 feet above the finished floor, except when protected.

Single sections of track must have at least two supports. Longer runs should be supported at least every 4 feet.

Track lighting must be securely grounded throughout its entire length.

Chapter Questions

1. What exactly is a lamp?

2. A fixture mounted on a concrete wall must be suitable for what type of location?

3. How far must recessed T-type fixtures be placed from thermal insulation?

4. What is "electric-discharge lighting"?

5. What prevents someone from installing too many fixture heads on a lighting track?

6. What weight of fixture requires independent support?

7. Are cold-storage warehouses considered wet locations?

8. The shell of a screw-shell lampholder must connect to what conductor of a standard 120-volt circuit?

9. Are splices allowed in lighting fixtures?

10. Can pull-chain fixtures be installed in the storage spaces of clothes closets?

CHAPTER
27
Appliances

The primary concern of this section of the Code is not the internal wiring of appliances, but the connection of these appliances to wiring systems.

The requirements for electrical appliances are covered in *Article 422* of the *NEC®*.

BRANCH CIRCUITS

A branch circuit that supplies an appliance must not be rated at less than the nameplate rating of the appliance it serves. (See *Chapter 18* of this text for requirements pertaining to branch circuits that supply more than one appliance.)

Central heating equipment must be supplied by an individual branch circuit. (Auxiliary equipment such as humidifiers, air filters, etc. can be connected to this circuit also.)

Branch circuits that supply storage-type water heaters must be rated at least 125 percent of the nameplate rating of the water heater.

Installation Methods

All appliances must be installed in an approved manner.

All exposed noncurrent-carrying metal parts of appliances must be grounded.

Flexible cord connections are allowed in the following circumstances:

1. To connect appliances that must be frequently interchanged, or that would be adversely affected by vibration.

2. Where ready removal for maintenance or repair is required. (In this case, the appliance must be identified as suitable for such use.)

Kitchen disposals in dwellings can be equipped with cord connections if:

1. The cord length is between 18 and 36 inches.
2. A grounding-type plug is used (unless the unit is double insulated and marked accordingly).
3. The receptacle is located so that the cord will not receive physical damage.
4. The receptacle is accessible.
5. Type S, SE, SEO, SO, SOO, ST, STO, STOO, SJ, SJE, SJEO, SJO, SJT, SJTO, SJTOO, SP-3, SPE-3, or SPT-3 3-wire cords are used.

Built-in dishwashers and trash compactors in dwellings can be equipped with cord connections if:

1. The cord length is between 3 and 4 feet.
2. A grounding-type plug is used (unless the unit is double insulated and marked accordingly).

3. The receptacle is located so that the cord will not receive physical damage.
4. The receptacle is accessible.
5. Type S, SE, SEO, SO, SOO, ST, STO, STOO, SJ, SJE, SJEO, SJO, SJT, SJTO, SJTOO, SP-3, SPE-3, or SPT-3 3-wire cords are used.

Electrically heated appliances must be located so as to provide protection between the appliance and any combustible materials.

Wall-mounted ovens and counter-mounted cooking units (note that these are *not* electric ranges, but wall ovens and cook-tops) can be either permanently connected (hardwired) or cord-and-plug connected. If cord-and-plug connected, the plug and receptacle combination cannot be considered a disconnecting means, and must be suitable for the temperatures it will encounter.

All appliances must have a means of disconnection. Any appliances fed by more than one source must have a disconnecting means for each source, and these disconnecting means must be grouped together and identified. The various disconnecting means allowed are as follows:

1. For permanently connected appliances of 300 watts (watts are also expressed as volt-amperes) or less, the branch-circuit overcurrent protective device (fuse or circuit breaker) is considered the disconnecting means.
2. For permanently connected appliances of greater than 300 watts, the branch-circuit switch or breaker can serve as the disconnecting means if it is within sight of the appliance or can be locked in the open position.
3. For cord connected appliances, a receptacle and plug or accessible connector can serve as the disconnecting means.
4. A cord and receptacle combination can be used as the disconnecting means for an electric range if the connection is accessible from the front of the range by removing a drawer.

5. An on/off switch in an appliance may be used as the disconnecting means, except in multifamily dwellings where the branch-circuit overcurrent device feeding the appliance is not on the same floor as the dwelling unit.

Switches on circuit breakers used as disconnecting means must be the indicating type ("ON" and "OFF" marked on them).

FIXED ELECTRIC SPACE HEATING EQUIPMENT

Fixed electric space heating equipment is covered in *Article 424* of the *NEC*®. These devices are among the most commonly used types of appliances. The equipment used for electric space heating can include heating cables, unit heaters, boilers, central systems, or other types. Process heating and room air-conditioning are not included in *Article 424* or the following requirements.

Electric space heaters (both built-in and portable) are used when quick, temporary heating is required. Because of the high cost of electric heating as opposed to gas or oil heat, however, electric heat is not very commonly used for central heating systems.

The requirements for electric space heating equipment are important because these devices have been the cause of a number of residential fires. While these fires have been due mostly to incorrect use, it has been nonetheless necessary to make these requirements more strict.

Branch Circuits

Branch circuits must be rated at least 125 percent of all heating and motor loads.

If the rating of the circuit does not correspond with a standard overcurrent protective device rating, the next higher size can be used.

Thermostats rated for continuous use at 100 percent of the rated load do not have to be sized to 125 percent of the circuit rating.

Installation Methods

Heaters to be installed in damp or wet locations must be specifically listed for such use, and must be installed so that water will not enter or accumulate in the unit.

Fixed electric space heating equipment may not be installed where it can be subjected to physical damage.

These heaters must be installed so that adequate space is given between the units and any combustible materials, unless specifically listed as usable in contact with combustible materials.

All noncurrent-carrying metal parts of fixed electric space heaters must be grounded.

An on/off switch in a fixed electric space heater may be used as the disconnecting means, except in multifamily dwellings where the branch-circuit overcurrent device feeding the appliance is not on the same floor as the dwelling unit.

A thermostat will be considered as both a controller and disconnecting means as long as it has a clearly marked "off" position, it opens all ungrounded conductors in the "off" position, it is within sight of the heater, and the system is designed so that it cannot be energized when the thermostat is in the "off" position.

Switches or circuit breakers used as disconnecting means must be the indicating type ("ON" and "OFF" marked on them).

Heaters having motor-compressors come under the authority of *Article 440*. (See *Chapter 30* of this text.)

ELECTRIC SPACE HEATING CABLES

Wiring above heated ceilings must not be more than 2 inches above the ceilings. The wires must have their ampacity calculated according to a 50°C ambient temperature. (The multipliers are shown in *Tables 310-16* through *310-19*.)

Heating cables cannot extend out of the room of their origination.

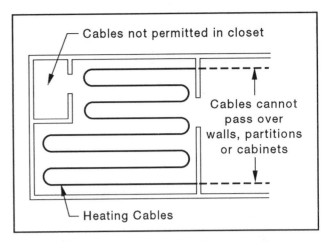

Fig. 27-1 Allowances for heating cables.

Heating cables cannot be installed above walls or partitions that extend to the ceiling (except isolated single embedded cables), in closets, or above cabinets that are closer to the ceiling than their depth. (See Figure 27-1.)

Special low-heat cables for humidity control are allowed to be installed in closets, but only in areas that are not over shelves.

Heating cables must be at least 8 inches from outlet boxes, and at least 2 inches from recessed fixtures, trims, or vents.

Embedded heating cables may be spliced only when necessary.

Heating cables cannot be installed in walls, except to go from one ceiling level to another.

Heating cables can be installed only on fire-resistant materials. Any exposed metal lathe must be covered with plaster before the heating cables are installed.

Runs of cables rated 2-3/4 watts per foot or less must be at least 1-1/2 inches apart (center to center). (See Figure 27-2.)

All of the heating cables and at least 3 inches of the nonheating leads must be embedded in plaster or dry board.

Cables must be secured at least every 16 inches (except cables specifically identified for different support spacings, but never more than every 6 feet).

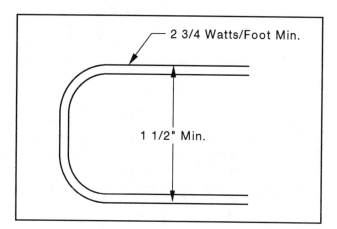

Fig. 27-2 Spacing of heating cables.

On drywall ceilings, the heating cables must be installed to the joists, and at least 2-1/2 inches on center. The cables may cross joists only when required by obstructions or at the ends of a room.

The nonheating leads for heating cables can be brought from the junction box to the ceiling by one of the following methods:

1. As single conductors in raceways.
2. As single or multiconductor Type UF, NMC, or MI cables.

Excess leads of heating cables should not be cut, but coiled in the ceiling, and then extended to the junction box only enough for 6 or 8 inches of lead to be free within the box.

In concrete or poured masonry floors, heating cables cannot exceed 16-1/2 watts per foot, and may not be placed closer than 1 inch apart on centers.

The cables must be secured by nonmetallic means while the concrete is being poured.

Cables may not be installed across expansion joints unless protected.

Heating cables (except grounded metal-clad cables) must be kept separated from metal in the concrete or masonry.

When leads leave the floor, they must be protected with bushings and use one of the following methods:

1. Rigid metal conduit.
2. Intermediate metal conduit.
3. Rigid nonmetallic conduit.
4. Electrical metallic tubing.

DUCT HEATERS

Heaters used in ducts must be identified as suitable for such use. (Units to be used within 4 feet of heat pumps need additional markings for such use on both the heater and the heat pump.)

Caution must be taken to ensure airflow through the heater according to the manufacturer's instructions.

Duct heaters must be accessible after installation, and be installed according to the manufacturer's instructions.

A disconnecting means must be installed within sight of the heater's controller.

ICE- AND SNOW-MELTING EQUIPMENT

Installation Methods

Cables may be installed no closer than 1 inch on center.

Cables must be installed on an asphalt or masonry base at least 2 inches thick, and must be covered by at least 1-1/2 inches of asphalt or masonry. Other investigated methods may also be used.

Cables must be secured while the masonry or asphalt is being applied.

When cables are installed across expansion joints, provision must be made for expansion and contraction.

Nonheating leads in masonry or asphalt must have a ground sheath or braid, or else have additional protection.

Nonheating leads enclosed in raceways must have between 1 and 6 inches. Bushings must be used to protect the cables as they leave the raceways. Rigid metal conduit, intermediate metal conduit, electrical metallic tubing, or other raceways may be used.

All noncurrent-carrying metal parts must be grounded.

Chapter Questions

1. Where are derating tables for electric ranges found?

2. Why can an on/off switch not be used as a disconnecting means in an apartment building that has the electrical panels in a central area?

3. Why are the required spacings of heating cables so important?

4. Which article of the Code covers appliances?

5. What parts of appliances must be grounded?

6. Can dishwashers have 5-foot cords?

7. Branch circuits for heating must be rated at what percentage of their load?

8. What must be done for heating cables that pass over expansion joints?

9. Where must the disconnecting means for a duct heater be installed?

CHAPTER
28

Cords and Cables

It is important to note that the *NEC*® does not consider flexible cords to be "wiring methods," but rather treats them as equipment connected to wiring systems. The most important thing about flexible cords is that they are to be used only in the proper locations and under the proper circumstances. Misuse of cords is one of the more hazardous types of Code violations.

Flexible cables are most commonly associated with industrial equipment, such as cranes and elevators. The requirements for these cables are very important for these installations, but are seldom needed by most electricians.

FLEXIBLE CORDS AND CABLES

(*Article 400*)

(See *Sections 400-4* and *400-5* for ampacity allowances for unusual cable types. This lengthy and rarely used section will not be included here.)

Where Permitted

As pendants.

For wiring fixtures.

To connect portable lamps or appliances. (Must be used with an attachment plug and fed from a receptacle outlet.)

As elevator cables.

To wire cranes and hoists.

To connect stationary equipment that must be frequently interchanged. (Must be used with an attachment plug and fed from a receptacle outlet.)

For connection of appliances identified for flexible cord connection. (Must be used with an attachment plug and fed from a receptacle outlet.)

As data processing cables.

For the connection of moving parts.

As temporary wiring.

Where Not Permitted

As a substitute for fixed wiring in a structure.

Where it must be run through holes in walls, ceilings, or floors.

Where it must be run through doors, walls, or similar openings.

Flexible cords may not be attached to building surfaces. (One connection for a tension take-up device within 6 feet of the cord termination is allowed.)

Flexible cords cannot be installed behind walls, ceilings, or floors.

Flexible cords cannot be installed in raceways, except when specifically allowed under specific articles of the Code.

Installation Methods

Splices are allowed in cords only when they are required for the repair of existing cord installations. In these cases, the splices may be made with splicing devices, brazing, welding, or soldering. (Soldered splices must be mechanically and electrically sound before the solder is applied.) The completed splice must match the insulation, outer sheath, and usage characteristics of the cable being spliced.

Cords must be connected to devices or fittings so that no tension will be applied to the wire connection terminals. The recommended methods for doing this are:
1. Knotting the cord.
2. Wrapping the cord with electrical tape.
3. Using fittings designed for this purpose.

In show windows or showcases (except when used for connecting chain-hung lighting fixtures, portable lamps, or merchandise being displayed), the following types of cords can be used: Type S, SO, SE, SEO, SOO, SJ, SJE, SJO, SJEO, SJOO, ST, STO, STOO, SJT, SJTO, SJTOO, or AFS.

When flexible cords pass through holes or openings in boxes, covers, or other enclosures, they must be protected by bushings or fittings.

PORTABLE CABLES OVER 600 VOLTS

All cables used on systems of more than 600 volts must contain conductors No. 8 AWG stranded copper or larger.

All cables operated at over 2000 volts must be shielded, and the shields must be grounded.

If connectors are used to connect lengths of cable, they must lock, and provisions must be made so that these connections will not be broken when the cables are energized. Tension must be eliminated at connections and terminations by appropriate methods.

Only permanent molded, vulcanized splices are allowed in these cables. Also, these splices must be accessible only by qualified personnel.

Chapter Questions

1. Are cords considered wiring or equipment?

2. Can cords or flexible cables be used in place of fixed wiring?

3. Can cords be run through doors?

4. Are splices allowed in cords?

5. Are cords ever allowed to be installed in raceways?

6. How must cords be protected when they pass through openings in boxes?

7. What kinds of splices can be used in cables over 600 volts?

CHAPTER
29
Motors

Obviously, *Article 430,* which covers electric motors, is one of the most important articles in the *NEC*®. This is not only because motors are so commonly used, but also because of their operating characteristics. The operation of electric motors involves not only current and voltage, but also magnetic fields and their associated characteristics.

At this point, it would be good to review basic motor and generator theory. Go back and reread "How Motors and Generators Work" in *Chapter 17* of this text. Having this knowledge fresh in your mind will help you in understanding *Article 430.*

The requirements of *Article 430* are the following:

ADJUSTABLE SPEED DRIVE SYSTEMS

The size of branch circuits or feeders to adjustable speed drive equipment must be based on the rated input current to the equipment.

If overload protection is accomplished by the system controller, no further overload protection is required.

The disconnecting means for adjustable speed drive systems can be installed in the incoming line, and must be rated at least 115 percent of the conversion unit's input current.

PART-WINDING MOTORS

If separate overload devices are used with standard part-winding motors, each half of the windings must be separately protected,

at one-half the trip current specified for a conventional motor of the same horsepower rating.

Each winding must have separate branch-circuit short-circuit and ground-fault protection, at no more than one-half the level required for a conventional motor of the same horsepower rating.

A single device (which has the one-half rating) can be used for both windings if it will allow the motor to start.

If a single time-delay fuse device is used for both windings, it can be rated no more than 150 percent of the motor's full-load current.

MOTOR AND AMPACITY RATINGS

All motors are considered to be continuous duty unless the characteristics of the equipment it drives ensure that the motor cannot operate under load continually.

Except for torque motors and AC adjustable voltage motors, the current rating of motors (used to determine conductor ampacities, switch ratings, and branch-circuit ratings) must be taken from *Tables 430-147* through *430-150*. These values may *not* be taken from a motor's nameplate rating (except for shaded-pole and permanent-split-capacitor fan or blower motors, which are rated according to their nameplates).

Separate overload protection for motors is to be taken from the motor's nameplate rating.

Multispeed motors must have the conductors to the line side of the controller rated according to the highest full-load current shown on the motor's nameplate (as long as each winding has its own overload protection, sized according to its own full-load current rating), and the conductors between the controller and the motor are to be based on the current for the winding supplied by the various conductors.

TORQUE MOTORS

The rated current for torque motors must be the locked-rotor current. This nameplate current must be used in determining branch-circuit ampacity, overload, and ground-fault protection.

ADJUSTABLE VOLTAGE AC MOTORS

The ampacity of switches and branch-circuit short-circuit and ground-fault protection for these motors must be based on the full-load current shown on the motor's nameplate. If no nameplate is present, these ratings must be calculated as no less than 150 percent of the values shown in *Tables 430-149* and *430-150*.

MOTOR LOCATIONS

In general, motors must be installed so that adequate ventilation is provided, and so that maintenance operations can be performed.

Open motors with commutators or collector rings must be located so that sparks from the motors cannot reach combustible materials. However, this does not prohibit installation of these motors on wooden floors.

Suitable enclosed motors must be used in areas where significant amounts of dust are present.

CONDUCTORS FOR MOTOR CIRCUITS

Proper sizing of motor conductors and overcurrent protection are the most important factors in a motor installation. Figure 29-1 is a brief outline of the various steps that must be taken in motor circuit design. These steps are explained in depth in this chapter; however, it is often easier to look at a brief outline, such as this one.

DESIGNING MOTOR CIRCUITS

For one motor:

1. Determine full-load current of motor(s) (*Table 430-150* for 3 phase).
2. Multiply full-load current × 1.25 to determine minimum conductor ampacity (*Art. 430-22a*).
3. Determine wire size (*Table 310-16*).
4. Determine conduit size. (*Tables 3A, B, & C*).
5. Determine minimum fuse or circuit breaker size (*Table 430-152*) (*Art. 240-6*).
6. Determine overload rating (*Art. 430-32*).

For more than one motor:

1. Perform steps 1 through 6 as shown above for each motor.
2. Add full-load current of all motors, plus 25% of the full-load current of the largest motor to determine minimum conductor ampacity (*Art. 430-24*).
3. Determine wire size (*Table 310-16*).
4. Determine conduit size. (*Tables 3A, B, & C*).
5. Add the fuse or circuit breaker size of the largest motor, plus the full-load currents of all other motors to determine the maximum fuse or circuit breaker size for the feeder (*Art. 430-62*) (*Art 240-6*).

Fig. 29-1 Design requirements of motor circuits.

Branch-circuit conductors that supply single motors must have an ampacity of at least 125 percent of the motor's full-load current rating, as shown in *Tables 430-147* through *430-150*. (This is necessary because motors cause temporary surges of current, which could overheat the conductors if they were not oversized.)

Motors used only for short cycles can have their branch-circuit ampacities reduced according to *Table 430-22(a), Exception*.

DC motors fed by single-phase rectifiers must have the ampacity of their conductors rated at 190 percent of the full-load current for half-wave systems, and 150 percent of the full-load current for full-wave systems. (This is because of the high levels of current that these motors can draw from the rectifiers.)

For phase converters, the single-phase conductors that supply the converter must have an ampacity of at least 2.16 times the full-load current of the motor or load being served. (This assumes that the voltages are equal. If they are not, the calculated current must be multiplied by the result of output voltage divided by input voltage.)

Conductors connecting secondaries of continuous-duty wound-rotor motors to their controllers must have an ampacity of at least 125 percent of the full-load secondary current.

When a resistor is installed separately from a controller, the ampacity of the conductors between the controller and the resistor must be sized according to *Table 430-23(c)*.

CONDUCTORS THAT SUPPLY SEVERAL MOTORS OR PHASE CONVERTERS

Conductors that supply two or more motors must have an ampacity of no less than the total of the full-load currents of all motors being served, plus 25 percent of the highest rated motor in the group. If interlock circuitry guarantees that all motors cannot be operated at the same time, the calculations can be made based on the largest group of motors that can be operated at any time.

Several motors can be connected to the same branch circuit if the following requirements are met:

1. Motors of 1 horsepower or less can be installed on general-purpose branch circuits without overload protection (assuming the installation complies with all other requirements).
2. The full-load current is not more than 6 amperes.
3. The branch-circuit protective device marked on any controller is not exceeded.

The conductors mentioned above must be provided with a protective device rated no greater than the highest rating of the protective device of any motor in the group, *plus* the sum of the full-load currents of the other motors.

Where heavy-capacity feeders are to be installed for future expansions, the rating of the feeder protective devices can be based on the ampacity of the feeder conductors.

Phase converters must have an ampacity of 1.73 times the full-load current rating of all motors being served plus 25 percent of the highest rated motor in the group. (This assumes that the voltages are equal. If they are not, the calculated current must be multiplied by the result of output voltage divided by input voltage.) If the ampere rating of the 3-phase output conductors is less than 58 percent of the rating of the single-phase input current ampacity, separate overcurrent protection must be provided within 10 feet of the phase converter.

CONDUCTORS THAT SUPPLY MOTORS AND OTHER LOADS

Conductors that supply motors and other loads must have their motor loads computed as specified above, and other loads computed according to their specific Code requirements, and the two loads added together.

If taps are to be made from feeder conductors, they must terminate into a branch-circuit protective device, and must:

1. Have the same ampacity as the feeder conductors.
      OR
2. Be enclosed by a raceway or in a controller, and be no longer than 10 feet.
      OR
3. Have an ampacity of at least one-third of the feeder ampacity, be protected, and be no longer than 25 feet.

In high-bay manufacturing buildings (more than 25 feet from floor to ceiling, measured at the walls), taps longer than 25 feet are permitted. In these cases:

1. The tap conductors must have an ampacity of at least one-third that of the feeder conductors.
2. The tap conductors must terminate in an appropriate circuit breaker or set of fuses.
3. The tap conductors must be protected from damage and installed in a raceway.
4. The tap conductors must be continuous with no splices.
5. The minimum size of tap conductors is No. 6 AWG copper or No. 4 AWG aluminum.
6. The tap conductors cannot penetrate floors, walls, or ceilings.
7. The tap conductors may be run no more than 25 feet horizontally, and no more than 100 feet overall.

Feeders that supply motors and lighting loads must be sized to carry the entire lighting load, plus the motor load.

OVERLOAD PROTECTION

Overload protection is not required where it might increase or cause a hazard, such as would be the case if overload protection were used on fire pumps.

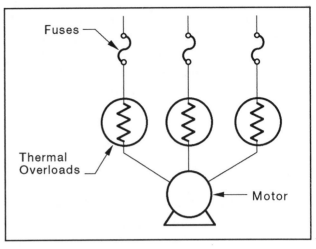

Fig. 29-2 Protection for motor circuits.

Continuous-duty motors of over 1 horsepower must have overload protection. This protection may be in one of the following forms:

1. An overload device that responds to motor current. These units must be set to trip at 115 percent of the motor's full-load current (based on the nameplate rating). Motors with a service factor of at least 1.15 or with a marked temperature rise of no more than 40°C can have their overloads set to trip at 125 percent of the full-load current. (See Figure 29-2.)
2. Any of several methods requiring overload protection built into motors (by the motor manufacturer, not by the installer).

Motors of 1 horsepower or less that are not permanently installed, are manually started, and are within sight of their controller are considered to be protected from overload by their branch-circuit protective device. They may be installed on 120-volt circuits up to 20 amps.

Motors of 1 horsepower or less that are permanently installed, automatically started, or not within sight of their controllers may be protected from overloads by one of the following methods:

1. An overload device that responds to motor current. These units must be set to trip at 115 percent of the motor's

full-load current (based on the name-plate rating). Motors with a service factor of at least 1.15 or with a marked temperature rise of no more than 40°C can have their overloads set to trip at 125 percent of the full-load current.

2. Any of several methods requiring overload protection built into motors (by the motor manufacturer, not by the installer).

3. Motors that have enough impedance to ensure that overheating is not a threat can be protected by their branch-circuit protective means only.

Wound-rotor secondaries are considered to be protected from overload by the motor overload protection.

Intermittent-duty motors can be protected from overload by their branch-circuit protective devices only, as long as the rating of the branch-circuit protective device does not exceed the rating specified in *Table 430-152.*

In cases where normal overload protection is too low to allow the motor to start, it may be increased to 130 percent of the motor's full-load rated current. Motors with a service factor of at least 1.15 or with a marked temperature rise of no more than 40°C can have their overloads set to trip at 140 percent of full-load current.

Manually started motors are allowed to have their overload protective devices momentarily cut out of the circuit during the starting period. The design of the cutout mechanism must ensure that it will not allow the overload protective devices to remain cut out of the circuit.

If fuses are used as overload protection, they must be installed in all ungrounded motor conductors, as well as in the grounded conductor for 3-phase, 3-wire systems that have one grounded wire (corner-ground delta systems).

If trip coils, relays, or thermal cutouts are used as overload protective devices, they must be installed according to *Table 430-*

37. The requirements for standard motor types are as follows:

1. Three-phase AC motors: One overload device must be placed in each phase.

2. Single-phase AC or DC, one wire grounded: One overload device, in the ungrounded conductor.

3. Single-phase AC or DC, ungrounded: One overload device, in either conductor.

4. Single-phase AC or DC, three wires, grounded neutral: One overload device, in either ungrounded conductor.

In general, overload protective devices should open enough ungrounded conductors to stop the operation of the motor.

When motors are installed on general-purpose branch circuits, their overload protection must be as follows:

1. Motors of over 1 horsepower can be installed on general-purpose branch circuits only when their full-load current is less than 6 amperes, they have overload protection, the branch-circuit protective device marked on any controller is not exceeded, and the overload device is approved for group installation.

2. Motors of 1 horsepower or less can be installed on general-purpose branch circuits without overload protection (assuming the installation complies with the other requirements mentioned above) if the full-load current is not more than 6 amperes, and the branch-circuit protective device marked on any controller is not exceeded.

3. When a motor is cord-and-plug connected, the rating of the plug and receptacle may not be greater than 15 amperes at 125 volts or 10 amperes at 250 volts. If the motor is over 1 horsepower, the overload protection must be built into the motor. The branch circuit must be rated according to the rating of the cord and plug.

4. The branch and overload protection must have enough time delay to allow the motor to start.

Overload protective devices that can restart a motor automatically after tripping are not allowed unless approved for use with a specific motor. This is *never* allowed if it can possibly cause injury to persons.

In cases where an instant shutdown of an overloaded motor would be dangerous to persons (as could be the case in a variety of industrial settings), a supervised alarm may be used, and an orderly rather than instant shutdown can be done.

SHORT-CIRCUIT AND GROUND-FAULT PROTECTION

Short-circuit and ground-fault protective devices must be capable of carrying the starting current of the motors they protect.

In general, protective devices must have a rating of no less than the values given in *Table 430-152*. When these values do not correspond with the standard ratings of overcurrent protective devices, the next higher setting can be used instead.

If the rating given in *Table 430-152* is not sufficient to allow for the motor's starting current, the following methods can be used:

1. A nontime-delay fuse of 600 amps or less can be increased enough to handle the starting current, but *not* to more than 400 percent of the motor's full-load current.
2. A time-delay fuse can be increased enough to handle the starting current, but *not* to more than 225 percent of the motor's full-load current.
3. The rating of an inverse time circuit breaker can be increased, but not to more than 400 percent of full-load currents that are 100 amps or less, or 300 percent of full-load currents over 100 amps.
4. The rating of an instantaneous trip circuit breaker can be increased, but

not to more than 1300 percent of the full-load current.
5. Fuses rated between 601 and 6000 amps can be increased, but not to more than 300 percent of the rated full-load current.

Instantaneous trip circuit breakers can be used as protective devices, but only if they are adjustable and part of a listed combination controller that has overload, short-circuit, and ground-fault protection in each conductor.

A motor short-circuit protector can be used as a protective device, but only when it is part of a listed combination controller that has overload, short-circuit, and ground-fault protection in each conductor, and does not operate at more than 1300 percent of full-load current.

For multispeed motors, a single short-circuit and ground-fault protective device can protect two motor windings, as long as the rating of the protective device is not greater than the highest possible rating for the smallest winding. (Multipliers of *Table 430-152* are used for the smallest winding's rating.)

A single short-circuit and ground-fault protective device can be used for multispeed motors, sized according to the full-load current of the highest rated winding. However, each winding must have its own overload protection, which must be sized according to each winding. Also, the branch-circuit conductors feeding each winding must be sized according to the full-load current of the highest winding.

If branch-circuit and ground-fault protection ratings are shown on motors or controllers, they must be followed, even if they are lower than Code requirements.

Fuses can be used instead of the devices mentioned in *Table 430-152* for adjustable speed drive systems, as long as a marking for replacements is provided next to the fuse holders.

For torque motors, the branch-circuit protection must be equal to the full-load current of the motor. If the full-load current is 800 amps or less and the rating does not correspond to a standard overcurrent protective device rating, the next higher rating can be used. If the full-load current is over 800 amps and different from a standard overcurrent device rating, the next *lower* rating must be used.

If the smallest motor on a circuit has adequate branch-circuit protection, additional loads or motors can be added to the circuit. However, each motor must have overload protection, and it must be ensured that the branch-circuit protective device will not open under the most stressing of normal conditions.

Two or more motors (each motor having its own overload protection) or other loads are also allowed to be connected to a single circuit in the following cases:

1. When the overload devices are factory installed, and the branch-circuit short-circuit and ground-fault protection are part of the factory assembly, or are specified on the equipment.
2. The branch-circuit protective device, motor controller, and overload devices are separate field-installed assemblies, are listed for this use, and are provided with instructions from the manufacturer.
3. All overload devices are marked as suitable for group installation, and marked with a maximum rating for fuses and/or circuit breakers. Each motor is marked as suitable for group installation, and marked with a maximum rating for fuses and/or circuit breakers. Each circuit breaker must be of the inverse time type, and must be listed for group installation.
4. The branch circuit must be protected with an inverse time circuit breaker rated for the highest rated motor and all other loads (including motor loads) connected to the circuit.
5. When the branch-circuit fuse or inverse time circuit breaker is not larger than allowed for the overload relay

that protects the smallest motor in the group. (See *Section 430-40.*)

For group installations, as described above, taps to single motors do not need branch-circuit protection in any of the following cases:

1. The conductors to the motor have an ampacity equal to or greater than the branch-circuit conductors.
2. The conductors to the motor are no longer than 25 feet, are protected, and have an ampacity at least one-third as great as the branch-circuit conductors.

MOTOR CONTROL CIRCUITS

Motor control circuits are circuits that turn the motor on or off, controlling the operation of the motor. A typical control circuit is shown in Figure 29–3. Note that when the circuit is energized, a set of contacts in parallel with the start switch make contact, keeping the circuit continuous, even when the start switch is no longer depressed.

Note also the overload contacts, which are in the control circuit. If any one of these contacts breaks, it will open the control circuit and cause the operation of the motor to cease.

The rules for motor control circuits are as follows:

Motor control circuits that are tapped from the load side of a motor's branch-circuit device and control the motor's operation are not considered branch circuits, and

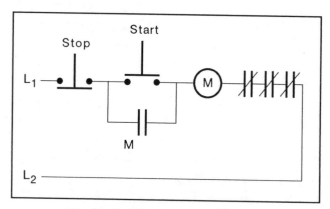

Fig. 29–3 Typical control circuit

can be protected by either a supplementary or branch-circuit protective device. Control circuits not tapped this way are considered signaling circuits, and must be protected accordingly. (See *Article 725.*)

Motor control conductors as described above must be protected (usually with an in-line fuse) in accordance with Column A of *Table 430-72(b),* except:

1. If they extend no further than the motor controller enclosure, they can be protected according to Column B of *Table 430-72(b)*.
2. If they extend further than the motor controller enclosure, they can be protected according to Column C of *Table 430-72(b)*.
3. Control circuit conductors taken from single-phase transformers that have only a 2-wire secondary are considered to be protected by the protection on the primary side of the transformer. However, the primary protection ampacity should not be more than the ampacity shown in *Table 430-72(b)* multiplied by the secondary-to-primary voltage ratio (secondary voltage divided by primary voltage).
4. When the opening of a control circuit would cause a hazardous situation (as would be the case with a fire pump, etc.), the control circuit can be tapped into the motor branch circuit with no further protection.

Control transformers must be protected according to *Article 450* or *Article 725,* except:

1. Control transformers that are an integral part of a motor controller and rated less than 50 volt-amperes can be protected by primary protective devices, impedance limiting means, or other means.
2. If the primary rating of the transformer is less than 2 amps, an overcurrent device rated at no more than 500 percent of the primary current can be used in the primary circuit.

3. By other approved means.
4. When the opening of a control circuit would cause a hazardous situation (as would be the case with a fire pump, etc.), protection can be omitted.

When damage to a control circuit would create a hazard, the control circuit must be protected (by raceway or other suitable means) outside of the control enclosure.

When one side of a motor control circuit is grounded, the circuit must be arranged so that an accidental ground will not start the motor.

Motor control circuits must be arranged so that they will be shut off from the current supply when the disconnecting means is in the open position.

MOTOR CONTROLLERS

Suitable controllers are required for all motors.

The branch-circuit protective device can be used as a controller for motors of 1/8 horsepower or less that are normally left running and cannot be damaged by overload or failure to start.

Portable motors of 1/3 horsepower or less may use a plug-and-cord connection as a controller.

Controllers must have horsepower ratings no lower than the horsepower rating of the motor they control, except:

1. Stationary motors 2 horsepower or less and 300 volts or less can use a general-use switch that has an ampere rating at least twice that of the motor it serves. General-use AC snap switches can be used on AC circuits to control a motor rated 2 horsepower or less and 300 volts or less and having an ampere rating of no more than 80 percent of the switch rating.
2. A branch-circuit inverse time circuit breaker rated in amperes only (no horsepower rating) can be used.

Unless a controller also functions as a disconnecting means, it does not have to open all conductors to the motor.

If power to a motor is supplied by a phase converter, the power must be controlled in such a way that, in the event of a power failure, power to the motor is cut off and cannot be reconnected until the phase converter is restarted.

Each motor must have its own controller, except when a group of motors (600 volts or less) uses a single controller rated at no less than the sum of all motors connected to the controller. This applies only in the following cases:

1. If a number of motors drive several parts of a single machine.
2. When a group of motors is protected by one overcurrent device, as specified elsewhere.
3. Where the group of motors is located in one room, and within sight of the controller.

A controller must be capable of stopping and starting the motor, and of interrupting its locked-rotor current.

The disconnecting means must be located within sight of the controller location and within sight of the motor, except in the following situations:

1. If the circuit is *over* 600 volts, the controller disconnecting means can be out of sight of the controller as long as the controller has a warning label that states the location of the disconnecting means to be locked in the open position.
2. One disconnecting means can be located next to a group of coordinated controllers on a multimotor continuous process machine.

The disconnecting means for motors 600 volts or less must be rated at least 115 percent of the full-load current of the motor being served.

A controller that operates motors over 600 volts must have the control circuit voltage marked on the controller.

Fault-current protection must be provided for each motor operating at over 600 volts. (See *Section 430-125[c]*.)

All exposed live parts must be protected. (See *Sections 430-132* and *430-133* if necessary.)

GROUNDING

The frames of portable motors that operate at more than 150 volts must be grounded or guarded.

The frames of stationary motors must be grounded (or permanently and effectively isolated from ground) in the following circumstances:

1. When supplied by metal-enclosed wiring.
2. In wet locations, when they are not isolated or guarded.
3. In hazardous locations.
4. If any terminal of the motor is over 150 volts to ground.

All controller enclosures must be grounded, except when attached to portable ungrounded equipment.

Controller-mounted devices must be grounded.

Chapter Questions

1. Why must motor conductors be oversized?

2. What are thermal overloads?

3. In what situations can thermal overloads by "cut out" of a motor circuit?

4. What article of the Code covers motors?

5. What is the required ampacity for branch-circuit conductors that supply a single motor?

6. Can intermittent-duty motors be protected from overloads by their branch-circuit devices only?

7. What table governs the installation of trip coils?

8. Overload protective devices should open which circuit conductors?

9. Is it ever allowed for motor overload devices *not* to shut down a motor?

10. Do taps to single motors require overload protection in all cases?

11. How far from a motor can its controller be mounted?

12. Do motors always require their own controllers?

13. Portable motors operating above what voltage must be grounded?

CHAPTER
30

HVAC Equipment

Not only are HVAC units one of the most common large electrical devices, but like motors they have some unique operating characteristics.

It is important to remember that in applying other sections of the *National Electrical Code*®, an air-conditioning or refrigeration system is considered a single machine.

Air-conditioning and refrigeration equipment is covered by *Article 440* of the *NEC*®. The requirements are as follows:

DISCONNECTING MEANS

A disconnect for a hermetic refrigerant motor-compressor must be sized at 115 percent of the nameplate current or branch-circuit current, whichever is higher.

Disconnect switches rated over 100 horsepower must be marked "DO NOT OPERATE UNDER LOAD."

A cord-and-plug connection can be considered as a disconnecting means for equipment such as room air conditioners, household refrigerators and freezers, water coolers, etc.

The disconnecting means must be located within sight of the equipment it serves. It may be located on or in the equipment.

BRANCH CIRCUITS

Branch-circuit ratings are marked on the nameplates of air-conditioning and refrigeration units (based on *Section 440-22*).

Branch-circuit conductors to a single motor-compressor must have an ampacity of at least 125 percent of the unit's rated current or the branch circuit's rated current, whichever is greater.

Conductors to more than one motor-compressor must be sized as follows:

1. The rated load or branch-circuit rating (whichever is larger) of the largest motor-compressor must be added to the full-load currents of all other motors or motor-compressors, plus 25 percent of the highest rated motor or motor-compressor in the group. Room air conditioners are excepted. In instances where control circuits are interlocked so that a second motor-compressor or a group of motor-compressors cannot be started when the largest is operating, the conductor size shall be determined from the rating(s) of the highest unit or units that can operate at any time.

When combination motor-compressor and lighting loads are connected to the same circuit, the conductor size must be based on the lighting load, plus the ampacity required for the motor-compressor(s). In instances where control circuits are interlocked so that motor-compressors and lighting loads cannot be operated at the same time, the conductor size shall be determined from the rating of either the motor-compressor or lighting load, whichever is higher.

CONTROLLERS AND OVERLOAD PROTECTION

Controllers and overload protection is virtually always built into these units. In the rare case that they are not, *Sections 440-41, 440-51, 440-52,* and *440-53* must be used. However, it is highly recommended to have an experienced equipment designer make these determinations.

ROOM AIR CONDITIONERS

All room air conditioners must be grounded.

The circuits supplying room air conditioners should be sized to at least 125 percent of the circuit size shown on the unit's nameplate (if no other equipment is supplied by the circuit). If other loads are also connected to the circuit, the circuit must be at least 200 percent of the unit's rating.

A plug and receptacle is allowed as the required disconnecting means for units of 250 volts or less, if the unit's controls are located within 6 feet of the floor, or a manual switch is located within sight of the unit in an accessible location.

Cords may not be longer than 6 feet for 208- or 240-volt units, and 10 feet for 120-volt units.

Chapter Questions

1. What article covers HVAC equipment?

2. What must be written on disconnects for HVAC equipment that uses disconnect switches rated over 100 horsepower?

3. Cord-and-plug connections can be considered disconnects for what types of equipment?

4. Branch-circuit conductors for a motor-compressor must be rated at what percentage of the unit's rated current?

5. What is the longest acceptable cord length for a 240-volt room air conditioner?

6. If a room air conditioner and another appliance are connected to the same circuit, what is the minimum rating for the circuit?

CHAPTER
31

Capacitors, Reactors, and Batteries

Capacitors, reactors, and batteries are very useful devices with special requirements, which are as follows:

CAPACITORS

(Article 460)

Capacitors are important components of AC circuits, and are especially useful in electronic circuits. Their basic function is to store and release relatively small amounts of electricity. The effect of capacitance is directly opposite to that of inductance. (Inductance is the effect caused by motors and other coils of wire.) They are often used in power wiring to reduce power factor losses caused by electric motors and other coils of wire. In these cases, a large capacitor is installed next to the motor that is causing the power factor problem, and connected in parallel with it.

Conductors

The ampacity of conductors for capacitor circuits must be at least 135 percent of the rated current of the capacitor being served. Conductors connecting capacitors to motors or motor circuits must be rated no less than one-third the full-load motor current, or 135 percent of the capacitor's rated current, whichever is greater.

Overcurrent protection must be provided in each ungrounded conductor in each capacitor installation. Overload devices are *not* required for conductors on the load side of motor overload protective devices. The rating of such devices should be as low as practical.

A disconnecting means must be provided in each ungrounded conductor in each capacitor installation. A disconnecting means is *not* required for conductors on the load side of motor overload protective devices. The rating of such a device may not be lower than 135 percent of the capacitor's rated current. The disconnecting means must open all ungrounded conductors, and may disconnect the capacitor from the line as a regular operating procedure.

Capacitor cases must be grounded.

RESISTORS AND REACTORS

(Article 470)

Reactors are coils of wire, properly called inductors. Only individual resistors and reactors are covered by this article, not components that are parts of other pieces of equipment.

Installation Methods

Resistors and reactors must not be located where they can be subjected to physical damage.

Resistors and reactors must either be placed 12 or more inches from combustible materials, or have a thermal barrier installed between them and any combustible materials.

Resistors or reactors operating at over 600 volts must be elevated or enclosed to avoid accidental contact with energized parts.

Resistors and reactors operating at over 600 volts must be placed at least 12 inches from combustible materials.

Reactor or resistor cases must be grounded.

STORAGE BATTERIES

(*Article 480*)

Storage battery installations must be carefully designed. Do *not* let your familiarity with them (such as with automobile batteries) lead you to think that storage batteries are relatively harmless. They can be explosive and very dangerous. Do *not* install storage batteries without a proper design layout. Remember, most design requirements are not shown here.

Installation Methods

Wiring to and from storage batteries must conform with all applicable parts of the *NEC®*.

Racks required for mounting storage batteries must be made of treated metal (that will resist corrosion from the electrolyte in the batteries) with nonconductive parts directly supporting the batteries, or made of fiberglass or the like.

Trays fitting into battery racks must be made of material that will resist the deteriorating action of the electrolyte.

Batteries must be installed in locations where they are well ventilated and where live parts will be guarded.

Vented cells must be provided with flame arresters.

Sealed cells must be equipped with pressure-release vents.

Chapter Questions

1. The ampacity of capacitor circuits must be at least what percentage of the capacitor's rated current?

2. Are overload devices required for equipment on the load side of capacitors?

3. If resistors are installed 6 inches from combustible materials, what devices are required?

4. How are resistors operating at over 600 volts required to be mounted?

5. Can storage batteries be explosive?

6. What makes storage batteries corrosive?

7. Must battery installations be ventilated?

PART
6

Special Locations and Systems

CHAPTER
32

Hazardous Locations

Hazardous locations are covered in *Articles 500* through *517* of the *NEC®*. These articles should be well understood by anyone wiring such locations. These areas are dangerous for a number of reasons, and each has its own special requirements and hazards.

All electrical installations in hazardous locations are inherently dangerous. Do *not* perform installations without carefully engineered layouts. If you do not have first-rate instructions, *do not install the wiring!* The installation requirements in this chapter are given to assist in the installation process, *not* as a substitute for an engineered layout. This work can be dangerous — don't take chances.

The following requirements are found in *Articles 500* through *504* of the *NEC®*.

GENERAL CONSIDERATIONS

Locations are classified as hazardous depending on the nature of the chemicals, dust, or fibers that may be present, and also upon their concentrations in the various environments.

In determining classifications, each room or area is considered separately.

Intrinsically safe circuits can be installed in any hazardous location, but must be kept isolated from all other wiring systems that are not intrinsically safe.

All equipment installed in hazardous areas (also called *classified* areas) must be approved for the *specific area* in which it is installed, not just approved for hazardous locations in general.

The wiring requirements for one type of hazardous location cannot be substituted for the requirements of another type of location. They are not interchangeable.

Locknut-bushing or double-locknut connections are not considered adequate bonding methods for hazardous locations. Other methods must be used.

By using ingenuity in laying out wiring for hazardous locations, a great deal of the wiring can be located outside of the hazardous areas, entering into the hazardous areas only where necessary. This avoids a great deal of the cost and difficulty of these installations.

CLASS 1 LOCATIONS

Class 1 locations are areas where flammable gases or vapors are present in amounts great enough to produce explosive or ignitible mixtures.

Class 1, Division 1 locations are areas where flammable concentrations of gases or vapors may be present under normal operating conditions; or in which such gases are frequently present because of maintenance or leakage; or where a breakdown might cause such vapors or gases to be present. One such location is the area around dispensing valves for propane or other flammable gases.

Class 1, Division 2 locations are areas in which flammable liquids, vapors, or gases are handled or processed, but in which the liquids, vapors, or gases are normally contained in closed containers from which they will escape only in abnormal cases. Typical of these areas is the area around propane storage tanks.

Groups in Class 1 are determined by the ignition temperatures for the types of gases or vapors in them.

Wiring methods in Class 1, Division 1 areas may be any of the following:

1. Threaded rigid metal conduit.
2. Threaded intermediate metal conduit.
3. Type MI cable (using fittings suitable for the location).

All boxes, fittings, and joints must be threaded for conduit and cable connections, and must be explosionproof.

Threaded joints must be made up with at least five threads fully engaged.

Type MI cable must be carefully installed so that no tensile (pulling) force is placed on the cable connectors.

Flexible connections can be used only where necessary, and must be made with materials approved for Class 1 locations.

Seals must be provided within 18 inches of any enclosure housing equipment that can produce sparks (such as switches, relays, circuit breakers, etc.). (See Figures 32-1 and 32-2.)

Seals must also be provided where conduits enter or leave a Class 1, Division 1 area.

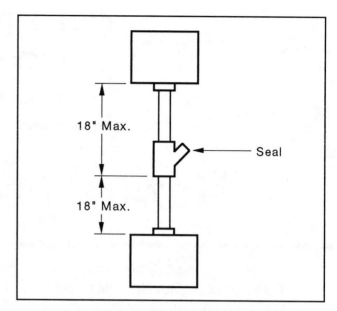

Fig. 32-1 Placement of seal on short run.

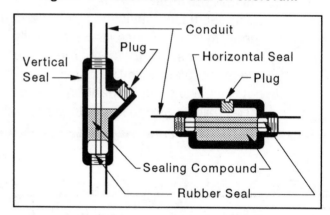

Fig. 32-2 Horizontal and vertical seals.

The seals may be on either side of the dividing wall. (See Figures 32-3 and 32-4.)

Wiring methods in Class 1, Division 2 areas may be any of the following:

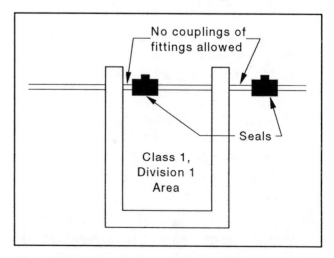

Fig. 32-3 Seals around Class 1, Division 1 areas.

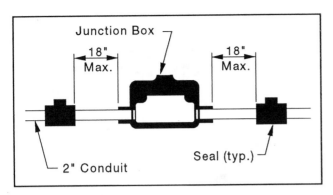

Fig. 32-4 Seals around junction boxes.

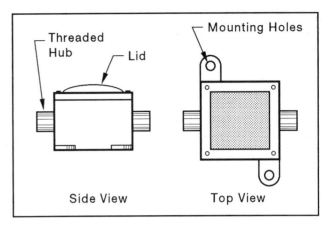

Fig. 32-5 Explosion proof box.

1. Threaded rigid metal conduit.
2. Threaded intermediate metal conduit.
3. Type MI, MV, MC, TC, or SNM cable (using fittings suitable for the location). These cables must be carefully installed so that no tensile (pulling) force is placed on the cable connectors.
4. Enclosed gasketed busway.
5. Enclosed gasketed wireway.
6. Type PLTC cable (installed under the provisions of *Article 725*).
7. Type MI, MV, MC, TC, SNM, and PLTC cables may be installed in cable trays.

All boxes, fittings, and joints must be threaded for conduit and cable connections, and must be explosionproof.

Threaded joints must be made up with at least five threads fully engaged.

Flexible connections can be used only where necessary, and must be made with materials approved for Class 1 locations.

CLASS 2 LOCATIONS

Class 2 locations are areas in which combustible dust may be present.

Class 2, Division 1 locations are areas in which the concentrations of flammable dust under normal conditions are sufficient to produce explosive or ignitible mixtures.

Class 2, Division 2 locations are areas in which flammable or ignitible dust is present, but normally not in quantities sufficient to produce a flammable or explosive mixture.

Wiring methods in Class 1, Division 1 areas may be any of the following:

1. Threaded rigid metal conduit.
2. Threaded intermediate metal conduit.
3. Type MI cable (using fittings suitable for the location).

All boxes, fittings, and joints must be threaded for conduit and cable connections, and must be approved for Class 2 locations. (See Figure 32-5.)

Type MI cable must be carefully installed so that no tensile (pulling) force is placed on the cable connectors.

Flexible connections can be made by any of the following means:

1. Liquidtight flexible metal conduit (with approved fittings).
2. Liquidtight flexible nonmetallic conduit (with approved fittings).
3. Extra-hard usage cord, with bushed fittings and dust seals.

When raceways extend between Class 2, Division 1 locations and unclassified locations, sealing may be done in any of the following ways (see Figure 32-6):

1. With raceway seals.
2. With a 10-foot horizontal run of raceway.
3. With a 5-foot vertical raceway.

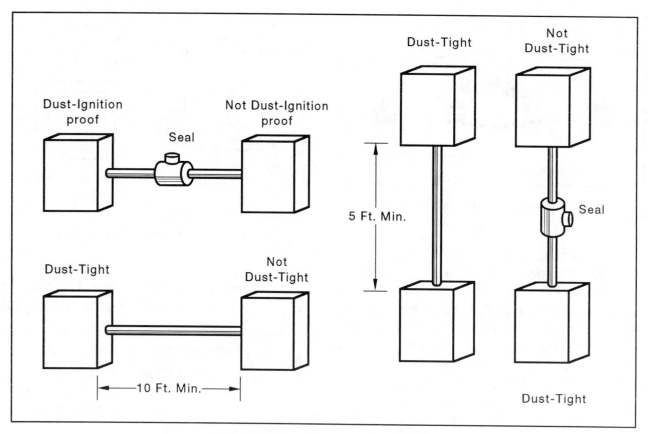

Fig. 32–6 Requirements for seals.

Wiring methods in Class 2, Division 2 locations may be any of the following:

1. Threaded rigid metal conduit.
2. Threaded intermediate metal conduit.
3. Type MI, MC, or SNM cable (using fittings suitable for the location). These cables must be carefully installed so that no tensile (pulling) force is placed on the cable connectors.
4. Enclosed gasketed busway.
5. Enclosed gasketed wireway.
6. Types MC, TC, and PLTC cables may be installed in cable trays.

When raceways extend between Class 2, Division 2 locations and unclassified locations, sealing may be done in any of the following ways:

1. With raceway seals.
2. With a 10-foot horizontal run of raceway.
3. With a 5-foot vertical raceway.

Flexible connections can be made by any of the following means:

1. Liquidtight flexible metal conduit (with approved fittings).
2. Liquidtight flexible nonmetallic conduit (with approved fittings).
3. Extra-hard usage cord, with bushed fittings and dust seals.

CLASS 3 LOCATIONS

Class 3 locations are termed hazardous areas only because of the presence of easily ignitible fibers or flyings, but these fibers or flyings are not likely to be suspended in the air in quantities sufficient to cause ignitible mixtures.

Class 3, Division 1 locations are areas where easily ignitible fibers or materials producing easily ignitible flyings are handled, manufactured, or used.

Class 3, Division 2 locations are areas where ignitible fibers are stored or handled.

Wiring methods in Class 3, Division 1 or 2 locations may be any of the following:

1. Threaded rigid metal conduit.
2. Threaded intermediate metal conduit.
3. Type MI, MC, or SNM cable (using fittings suitable for the location). These cables must be carefully installed so that no tensile (pulling) force is placed on the cable connectors.
4. Dusttight wireway.

Flexible connections can be made by any of the following means:

1. Liquidtight flexible metal conduit (with approved fittings).
2. Liquidtight flexible nonmetallic conduit (with approved fittings).
3. Extra-hard usage cord, with bushed fittings and dust seals.

INTRINSICALLY SAFE SYSTEMS

Intrinsically safe systems must be designed for the specific installation, complete with control drawings. This is required by the *NEC*®, and no intrinsically safe system installation may be attempted without such drawings.

COMMERCIAL GARAGES

Commercial garages are hazardous areas, although not all areas of a garage are considered to be so. The requirements for such garages are shown in *Article 511,* and are as follows:

In commercial garages, the entire area of the garage, from the floor up to a height of 18 inches, must be considered a Class 1, Division 2 location. (See Figure 32-7.)

Any pit or depression in the floor shall be considered a Class 1, Division 1 location from the floor level down. (See Figures 32-8 and 32-9.)

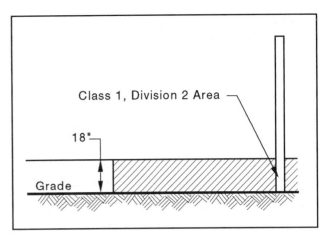

Fig. 32-7 Definition of Class 1, Division 2 area.

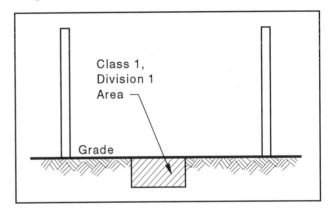

Fig. 32-8 Definition of Class 1, Division 1 area.

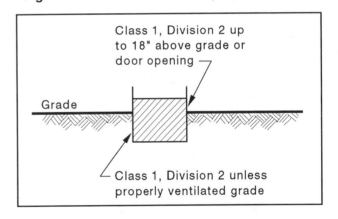

Fig. 32-9 Class 1 locations.

Adjacent areas such as storerooms, switchboard rooms, etc. are not considered to be classified areas if they have ventilating systems that provide four or more changes per hour, or if they are well separated from the garage area by walls or partitions.

Areas around fuel pumps are covered by *Article 514.*

Wiring methods in areas above Class 1 locations may be any of the following:

1. Rigid metallic conduit.
2. Intermediate metal conduit.
3. Rigid nonmetallic conduit.
4. Electrical metallic tubing.
5. Type MI, TC, SNM, or MC cable.

Plug receptacles above Class 1 locations must be approved for the purpose.

Electrical equipment that could cause sparks located above Class 1 locations must be totally enclosed, or located at least 12 feet from the floor.

All receptacles installed where hand tools, diagnostic equipment, or portable lighting devices are to be used must have ground-fault protection.

AIRCRAFT HANGARS

The requirements for aircraft hangars are shown in *Article 513* of the Code. It is important that in addition to concerns over volatile fuels, special care must also be taken with aircraft because of problems with static electricity. Such static can cause sparks that, in turn, can ignite volatile substances (like gasoline). This requires special attention to grounding systems in these areas.

The requirements of *Article 513* are as follows:

In aircraft hangars, the entire area, from the floor up to a height of 18 inches, must be considered a Class 1, Division 2 location. (See Figure 32-10.)

Any pit or depression in the floor shall be considered a Class 1, Division 1 location from the floor level down.

Adjacent areas such as storerooms, switchboard rooms, etc. are not considered to be classified areas if they are well separated from the hangar itself by walls or partitions.

Wiring methods in areas above Class 1 locations may be any of the following:

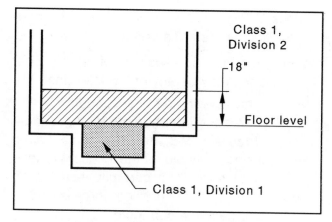

Fig. 32-10 Class 1 locations.

1. Rigid metallic conduit.
2. Intermediate metal conduit.
3. Electrical metallic tubing.
4. Type MI, TC, SNM, or MC cable.

Electrical equipment that could cause sparks located above Class 1 locations must be totally enclosed, or located at least 10 feet from the floor.

Aircraft electrical systems must be deenergized when an aircraft is stored in a hangar and, whenever possible, when the aircraft is being serviced.

Aircraft batteries cannot be charged when they are installed in an aircraft located partially or fully inside a hangar.

SERVICE STATIONS

Gasoline dispensing and service stations are hazardous locations, and as stated in the beginning of this chapter, there are dangers involved with wiring in these locations. No installations should be done in these areas without carefully engineered drawings. The requirements shown here are for informational purposes and do not replace a properly engineered layout.

The requirements for service stations are shown in *Article 514* of the Code and are as follows (see Figures 32-11 through 32-14):

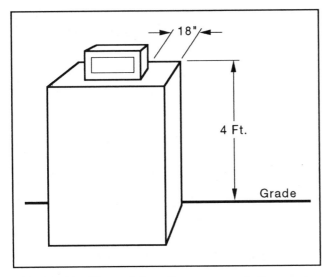

Fig. 32-11 Class 1 locations around dispensing locations.

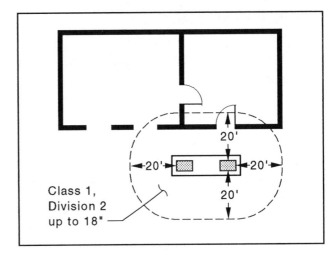

Fig. 32-12 Class 1, Division 2 locations.

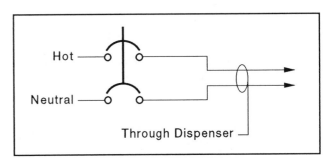

Fig. 32-13 Location of circuit breaker in dispensing locations.

Installation Methods

Table 514-2 specifies which locations in service stations are considered classified. Wiring in these areas must be in accordance with their class and division.

Underground wiring must be in either threaded rigid metal conduit or threaded intermediate metal conduit. Any portion under Class 1, Division 2 areas must be considered Class 1, Division 1 locations, and shall be considered as such up to the point at which the raceway emerges from the ground or floor. Properly installed Type MI cable is also permitted. Also, rigid non-metallic conduit can be used if installed at least 2 feet below grade, and if rigid metal conduit is used for the last 2 feet of the run prior to its emergence from the ground or floor.

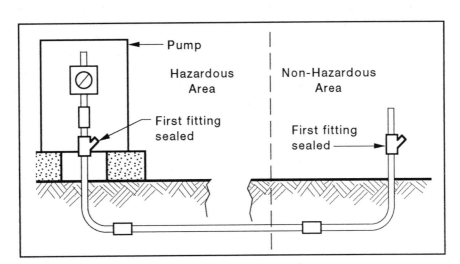

Fig. 32-14 Definitions of hazardous areas.

BULK STORAGE PLANTS

The requirements for bulk storage facilities are found in *Article 515* of the *NEC*® and are as follows:

Installation Methods

Table 515-2 specifies which locations in bulk storage plants are considered classified. Wiring in these areas must be in accordance with their class and division.

Underground wiring must be in either threaded rigid metal conduit or threaded intermediate metal conduit. Any portion under Class 1, Division 2 areas must be considered Class 1, Division 1 locations, and shall be considered as such up to the point at which the raceway emerges from the ground or floor. Properly installed Type MI cable is also permitted. Also, rigid non-metallic conduit can be used if installed at least 2 feet below grade, and if rigid metal conduit is used for the last 2 feet of the run prior to its emergence from the ground or floor. (See Figures 32-15 through 32-20.)

All wiring above Class 1 locations must be done by one of the following methods:

1. Rigid metal conduit.
2. Intermediate metal conduit.
3. Electrical metallic tubing.

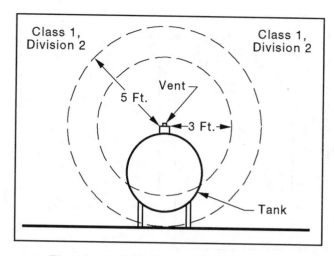

Fig. 32-15 Definition of Class 1 areas.

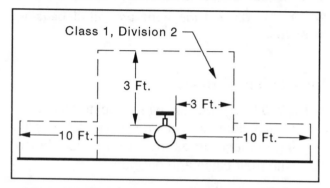

Fig. 32-16 Definition of Class 1 areas.

4. Schedule 80 PVC conduit.
5. Type MI, TC, SNM, or MC cable.

Equipment above Class 1 locations must be totally enclosed.

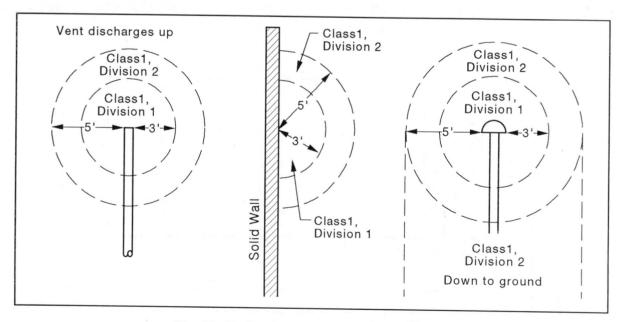

Fig. 32-17 Class 1 areas around vent pipes.

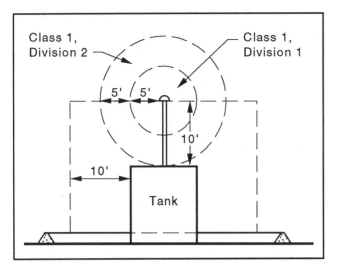

Fig. 32-18 Class 1 areas around tanks.

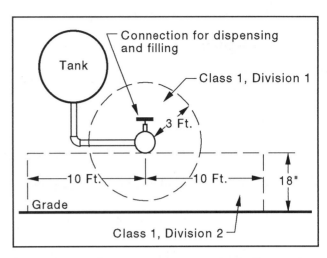

Fig. 32-19   Class 1 areas around dispensing locations.

SPRAY AREAS

Spray areas are very hazardous due to the high levels of solvents used for spray processes. (See Figure 32-21.)

The special requirements for spray areas are found in *Article 516* of the *NEC®* and are the following:

Installation Methods

Figures 1, 2, 3, and *4* of *Article 516* specify which locations in spray areas are considered classified. Wiring in these areas must be in accordance with their class and division.

Adjacent areas such as storerooms, switchboard rooms, etc. are not considered to be classified areas if they are well separated from the spray area by tight walls or partitions without communicating openings.

All wiring above Class 1 locations must be done by one of the following methods:

1. Rigid metal conduit.
2. Intermediate metal conduit.
3. Electrical metallic tubing.
4. Rigid nonmetallic conduit.
5. Type MI, MC, TC, or SNM cable.

Equipment above Class 1 locations must be totally enclosed.

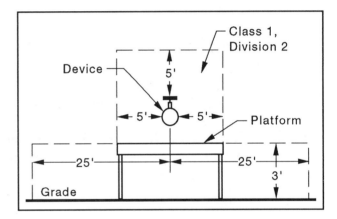

Fig. 32-20 Class 1 locations.

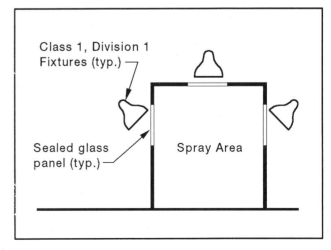

Fig. 32-21 Lighting for spray areas.

Chapter Questions

1. Why are conduits sealed before entering hazardous locations?

2. What is an explosionproof fitting?

3. What parts of commercial garages are considered hazardous?

4. What special hazards exist in aircraft hangars?

5. What is a bulk storage plant?

6. Why are engineered layouts important for hazardous areas?

7. What are Class 1 areas?

8. What are Class 2 areas?

9. What are Class 3 areas?

10. How many threads deep must a threaded connection be made in hazardous areas?

11. Which article contains the requirements for commercial garages?

12. Electrical devices that can cause sparks must be located how far above the floor above Class 1 locations in aircraft hangars?

CHAPTER
33

Health Care Facilities

It is very important to wire all health care facilities exactly according to carefully prepared and engineered drawings, not only for safety's sake, but also because of the very exacting requirements of the Health and Rehabilitation Services administration (HRS). HRS has their own set of requirements and perform their own inspections. *Never* begin an installation in a health care facility without engineered drawings. The HRS is *very* serious about their requirements, and I have seen instances when electrical contractors had to bring their work up to HRS's standards at costs of up to $7 million—without reimbursement.

The extensive requirements for health care facilities are found in *Article 517* of the *NEC®*. As you go through this article, pay special attention to *Section 517-2,* which defines many of the sometimes confusing terms used in this article.

This article is extremely important, since health care facilities can have not only hazardous areas within them, but also large numbers of people, many of whom are not mobile (or *ambulatory*, in medical lingo) and cannot leave the area quickly in the event of a hazardous equipment failure. Hospital wiring *must* be safe—many people's lives are entirely dependent upon it.

The installation requirements of *Article 517* are the focus of this chapter. You will note in your review of this article that there are many other requirements, covering engineering and design. I have omitted them from discussion here, since they would only serve to confuse you at this point.

The installation requirements are as follows:

GENERAL AREAS

General areas include general health care areas, including such areas as a doctor's examining room in a multifunction building. These areas do not include business offices, waiting rooms, or patient rooms in nursing homes.

Except where specifically prohibited (which is frequently), any applicable types of wiring can be used in these areas.

In patient care areas, all exposed metal surfaces and grounding terminals of receptacles must be grounded with an insulated copper grounding conductor. This conductor must be sized according to *Table 250-95.*

Wiring methods in patient care areas must be one or more of the following:

1. Rigid metal conduit.
2. Intermediate metal conduit.
3. Electrical metallic tubing.
4. Type MI, MC, or AC cable.

Equipment grounding buses in panelboards for essential and normal branch circuits must be bonded with at least a No. 10 AWG insulated copper conductor.

Ground-fault protection is required in patient care areas, for both branch circuits and feeders.

Bed locations must be provided with at least two branch circuits; one from the emergency system, and one from the normal system.

The emergency circuit must supply at least one receptacle for that bed location only. The receptacle(s) must be identified, as also must be the panelboard and circuit number.

All circuits on the normal power must be from a single panelboard.

All circuits on the emergency power must be from a single panelboard.

All patient bed locations must have at least six hospital grade receptacles. All these receptacles must be bonded to an equipment grounding point with a bonding jumper of at least No. 10 AWG insulated copper.

All panelboards must be grounded by one of the following means:

1. Where a locknut-bushing connection is used, grounding bushings and continuous copper bonding jumpers must be used. (Size jumper according to *Table 250-95.*)
2. By connecting feeder raceways or cables to threaded hubs on the terminating enclosures.
3. Other approved devices such as bonding locknuts, etc.

Special design requirements for critical care areas, essential electrical systems, anesthetizing locations, X-ray locations, signaling systems, and isolated power systems are not shown here.

Chapter Questions

1. Which article of the *NEC* ®covers health care facilities?

2. Are doctor's examination rooms considered to be health care facilities?

3. Can rigid nonmetallic conduit be used in patient care areas?

4. What special type of protection is required for feeders to patient care areas?

5. What is the minimum number of receptacles that can be installed at patient bed locations?

6. What grade of wiring device must be used in these areas?

7. What type of bonding jumper is required for these receptacles?

8. Are threaded hubs an acceptable means of grounding panelboards in health care facilities?

CHAPTER
34

Theaters and Places of Assembly

As with health care facilities, coverage of these articles of the Code is focused on the installation requirements of the Code, rather than on the lengthy engineering and design requirements.

Since many people gather in these locations, there is a tremendous capacity for destruction; therefore, the concerns here are those of safety. The Iroquois Theater Fire in Chicago killed more than 600 persons, and is a big reason that the *National Electrical Code*®became so widespread.

PLACES OF ASSEMBLY

(Article 518)

Places of assembly are buildings or parts of buildings intended for the gathering together of 100 or more persons.

Temporary wiring in these areas must conform to *Article 305* (see *Chapter 8* of this text), except that cords can be laid on floors if they are protected from contact by the public.

Acceptable wiring methods in places of assembly are the following:

1. Rigid metal conduit.
2. Intermediate metal conduit.
3. Electrical metallic tubing.
4. Rigid nonmetallic conduit and other nonmetallic raceways, if encased in 2 or more inches of concrete.
5. Type MI or MC cable.

The number of conductors in raceways is the same for places of assembly as for all standard types of wiring, with one exception: The limit of thirty current-carrying conductors in an auxiliary gutter or wireway does not apply.

THEATERS

(Article 520)

Temporary wiring in these areas must conform to *Article 305* (see *chapter 8* of this text), except that cords can be laid on floors if they are protected from contact by the public.

Acceptable wiring methods in theaters are the following:

1. Rigid metal conduit.
2. Intermediate metal conduit.
3. Electrical metallic tubing.

4. Rigid nonmetallic conduit and other non-metallic raceways, if encased in 2 or more inches of concrete.
5. Type MI or MC cable.

The number of conductors in raceways is the same for theaters as for all standard types of wiring, with one exception: The limit of thirty current-carrying conductors in an auxiliary gutter or wireway does not apply.

All live parts (including parts of lighting switchboards) must be protected from accidental contact by persons or objects.

Circuits for footlights, border lights, and proscenium sidelights cannot have an ampacity of greater than 20 amps, except when heavy-duty lampholders are used. Conductors to such lighting must have insulation capable of withstanding the heat encountered. The conductor insulation may not be rated less than 125°C.

Requirements for specific types of stage equipment are not shown here.

MOTION PICTURE AND TELEVISION STUDIOS

(*Article 530*)

Permanent wiring in these areas must be either Type MC or MI cable, or an approved raceway.

Portable wiring in these areas must be done with approved cord or cables. Splices are allowed if done by approved methods, and if the overcurrent protection to the cables in question is no more than 20 amperes.

Switches used for stage set lighting must be externally operable.

Receptacles for DC plugging boxes must be rated at least 30 amperes.

All live parts must be protected from accidental contact by persons or objects.

Overcurrent protection levels for various types of equipment in motion picture and television studios are as follows:

1. Stage cables may be protected at up to 400 percent of their rated ampacity.
2. Feeders (that go only from substations to stages, and operate only for periods of 20 minutes or less) can be protected for up to 400 percent of their rated ampacity.

Feeders to studios can be sized based on the derating factors of *Table 530-19(a)*.

All noncurrent-carrying parts of all equipment must be grounded.

Enclosed and gasketed lighting fixtures must be used in film storage areas. No other equipment is allowed in these areas.

Switches controlling the lights in film storage areas must be located outside the areas, must disconnect all ungrounded conductors (no standard 3-way switches), and must have a pilot light that indicates when the lights in the storage area are on.

Requirements for projectors are omitted here.

ELECTRIC SIGNS

(*Article 600*)

Installation Methods

Each outline lighting installation or sign (other than portable signs) must be controlled with an externally operated switch or circuit breaker, which must open all ungrounded conductors.

The disconnecting means mentioned above must be within sight of the sign it protects. If the sign has an electronic or electromechanical controller not located in the sign, the disconnecting means can be installed within sight of the controller, rather than within sight of the sign. (See Figure 34–1.)

Control devices that control transformers must be rated for the inductive load, or have an amperage rating at least twice that of the transformer.

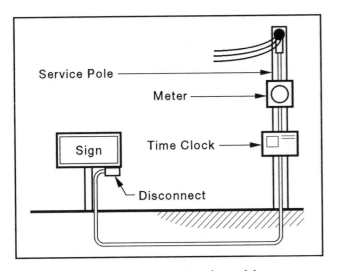

Fig. 34-1 Diagram for sign wiring.

Wiring to signs must terminate in junction boxes.

Electric signs must be listed.

All metal parts of signs (except isolated parts) must be grounded.

Circuits to lamps, ballasts, and transformers cannot be rated at more than 20 amperes.

Circuits that feed electric-discharge lighting transformers cannot be rated at more than 30 amperes.

All commercial buildings accessible to the public must have at least one sign outlet on its own 20-amp circuit.

Flashers, cutouts, etc. must be installed in metal boxes that can be accessed after the sign is installed.

A weatherproof receptacle and attachment plug must be provided for each sign or independent section of a sign.

Cord types S, SJ, SJO, SJTO, SO, or STO (3-wire) can be used for connecting the above-mentioned attachment plugs.

Signs must be installed so that their lowest part is at least 16 feet above areas open to vehicular traffic. (They can be mounted lower if suitably protected.)

Outdoor portable signs must be ground-fault protected. The protective device can be in

either the sign or the power system that supplies the sign.

Any normal wiring method can be used for supplying signs operating at 600 volts or less.

Requirements for the manufacturing of electric signs are not shown here.

MANUFACTURED WIRING SYSTEMS

(*Article 604*)

Installation Methods

Manufactured wiring systems can be installed in dry, accessible locations, and (when listed for this use) in plenums and spaces for environmental air.

One end of a cable an be extended into hollow walls where necessary to reach a switch or outlet.

Manufactured wiring systems are made from Type AC or MC cable, and are subject to any restrictions placed on these cable types.

All unused outlets must be capped.

The entire system must be properly grounded.

OFFICE FURNISHINGS

(*Article 605*)

Installation Methods

Wiring in office furnishings (usually partitions) must be of a type for which the furnishings are listed.

All wiring and connections must be done in the wiring channels of the office furnishings provided for this use.

Electrical connections between partitions must be made with types of wiring listed for these partitions; or with flexible cords if all of the following conditions are met:

1. Extra-hard usage cord must be used.

2. The partitions must be mechanically secured to each other and continuous.
3. The cord must be only as long as necessary, but never longer than 2 feet.
4. The cord must terminate in an attachment plug and cord-connector with strain relief.

Any lighting used with such partitions must be listed for this purpose.

Where lighting such as that mentioned above uses cord-and-plug connections, the conductors in the cord must be at least No. 18 AWG copper, and the cords cannot be longer than 9 feet.

Fixed partitions must be permanently connected to building wiring systems.

Free-standing partitions can be connected to building electrical systems, but are not required to be so connected.

Free-standing partitions that are mechanically continuous and no more than 30 feet long can be connected to the building electrical system with a flexible cord. Such a cord must be no more than 2 feet long, and must be an extra-hard usage-type cord with conductors no smaller than No. 12. The cord must have an insulated grounding conductor.

A receptacle supplying power to a partition as that mentioned above must be on a separate circuit with no other loads connected to it, and within 12 inches of the partition(s) it serves.

Groups of partitions may not have more than thirteen receptacles in them.

Partitions or groups of partitions are not allowed to contain multiwire circuits (such as a 3-phase, 4-wire network).

Chapter Questions

1. When can rigid nonmetallic conduit be used in theaters?

2. Can more than thirty conductors be installed in wireways in theaters?

3. When can footlight circuits in theaters be rated at more than 20 amps?

4. Can a 30-amp cable in a motion picture studio be spliced?

5. Can 3-way switching be used in film storage areas?

6. Do electric signs need to be listed?

7. What special protection do portable outdoor signs require?

8. Can manufactured wiring systems be installed in plenums?

9. What types of cords must be used between office furnishing partitions?

10. Can a group of office furnishing partitions contain twelve receptacles?

CHAPTER
35

Mobile Homes and RV Parks

The whole of *Article 550* of the *National Electrical Code*®covers mobile homes. In this text, however, I have omitted the study of the numerous sections that contain only requirements for the manufacture of mobile homes. The electrician must take care in interpreting this and similar articles not to confuse the requirements for building mobile homes in a factory with those for wiring the mobile home after it has left the factory.

MOBILE HOMES

Proper wiring requirements for mobile homes have been a hotly debated topic among code panelists and electrical inspectors for a long time. As of the 1993 edition of the *NEC*®, the key points to remember about mobile homes are the following:

First, mobile homes are not considered permanent residences unless they are installed upon a permanent foundation. If so, they cannot be wired according to *Article 550*.

Next (and very important) is that the circuit breaker or fuse panel inside the mobile home (or occasionally mounted on an outside wall) is *not* a service panel but a feeder panel. The service equipment is usually mounted on a pole or some other mounting structure within 30 feet of the mobile home. The panel inside the mobile home is a feeder panel; it must not have a bonding jumper installed, and must have the neutral isolated from equipment grounding. This requires the use of 4-wire cables to appliances such as clothes dryers and ranges (two hot

conductors, one neutral, and one equipment grounding conductor).

Supply cords to the mobile home must be between 21 and 36-1/2 feet long. (No, I do not know why the Code writers chose 36-1/2 feet.)

All metal parts in the mobile home, both electrical and not electrical, must be connected to a grounding bus in the distribution panel. This includes the chassis. They may *not* be connected to the neutral bus.

I will not be covering all of the definitions shown in *Section 550-2,* but you should study them carefully; after all, if you do not really understand the terminology used, you will not be able to properly understand the requirements either.

Service Equipment

Services for mobile homes must meet the basic requirements for electrical services that are given in *Article 230*. In addition to the requirements of *Article 230, Article 550* sets forth

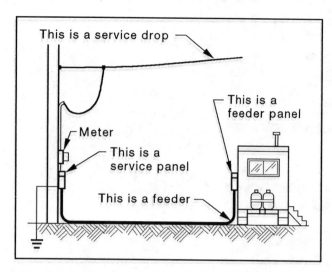

Fig. 35-1 Service and feeder for mobile homes.

certain other requirements that take precedence over the requirements of *Article 230* in cases of conflict. Some of the requirements of this article are more stringent than those of *Article 230,* and some are more lenient, due to the different circumstances surrounding mobile homes and mobile home parks. The requirements are as follows:

Service equipment for mobile homes must be located outside (not in or on) the mobile home. (See Figure 35-1.) It must be located within sight, and not more than 30 feet away, unless all of the following requirements are met:

1. A disconnecting means suitable for use as service equipment is located no more than 30 feet from the mobile home it serves, and also within sight.
2. A grounding electrode (see *Article 250*) is present or installed at the disconnecting means.
3. A grounding electrode conductor connects the equipment grounding terminal of the disconnecting means to the grounding electrode conductor.

Service equipment for mobile homes cannot be rated less than 100 amperes.

The outdoor mobile home disconnecting means must be mounted so that the bottom of the enclosure is at least 2 feet above

grade, and the operating handle is no more than 6-1/2 feet above grade.

All mobile home services must be grounded.

Mobile home lot feeder conductors must be suitable for the loads they carry, and may never have ampacities less than 100 amperes.

Power Supply

The power supply to a mobile home consists of a 50-amp (40-amp if the mobile home has gas heating and cooking appliances) mobile home rated cord installed between the service equipment and the distribution panel inside the mobile home. *Section 550-5* gives the following requirements:

When its calculated load does not exceed 50 amperes, the power supply to a mobile home must be a 50-amp mobile home supply cord with an integral cap, or a permanently installed feeder. (Mobile homes with gas or oil-fired cooking and heating systems are allowed to use a 40-amp cord assembly.)

The cord mentioned above must be a 4-conductor cord, permanently connected to the distribution panel or to a junction box connected to the distribution panel.

The attachment cap plug must be a 50-amp, 4-wire, 3-pole, grounding-type cap.

The power-supply cord must be between 21 and 36-1/2 feet long. The distance between the plug-in location and the mobile home must be at least 20 feet. This cord must be marked as suitable for mobile homes, and must state its amperage rating.

The feeder assembly (cord or other type) must enter the mobile home through the floor, exterior wall, or roof. Where the cord passes through floors or walls, bushings or conduits must be used to protect the cord. Cords may be run through walls between the panel and the floor, but in these cases they must be installed in conduits having a size of at least 1-1/4 inches (trade size).

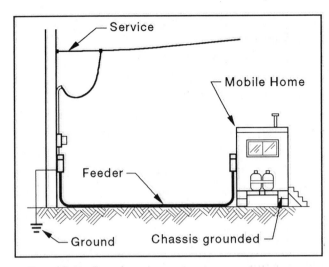

Fig. 35-2 Overhead service to a mobile home.

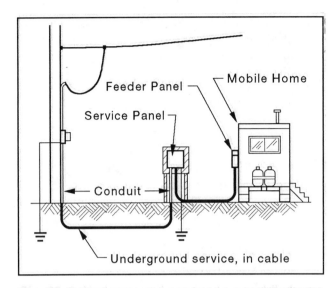

Fig. 35-3 Underground service to a mobile home.

In cases where the calculated load exceeds 50 amperes, or where permanently installed feeders are used for other reasons (which is more commonly the case), the mobile home can be supplied by one of the following means:

1. An overhead mast weatherhead service (see *Article 230*) with four insulated, color-coded conductors, one of which must be an equipment grounding (green) conductor. (See Figure 35-2.)
2. A metal raceway or rigid nonmetallic conduit from the disconnecting means in the mobile home to the underside of the mobile home. (See Figures 35-3 through 35-5.)

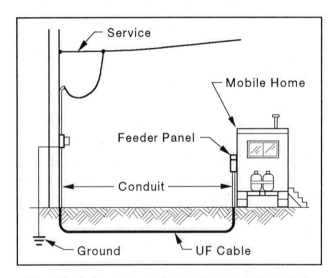

Fig. 35-4 Service equipment mounted on pole.

Receptacles

In general, receptacles in mobile homes must be installed according to the same requirements as for receptacles in conventional residences. But, as stated earlier, care must be taken so that range and dryer receptacles are wired not only with hots and a neutral, but also with a separate insulated grounding conductor (green wire).

The following requirements are found in *Article 210* (Branch Circuits) and *Section 410-L* (Receptacles, Cord Connectors, and Attachment Plugs), and apply to all mobile home installations:

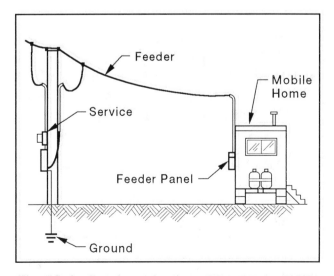

Fig. 35-5 Overhead feeder cable run to mobile home

A 15-amp multiple (not single) receptacle is allowed to be installed on a 20-amp laundry circuit.

All receptacles in mobile homes must be of the standard grounding type.

Any receptacles installed within 6 feet of a sink or lavatory must be protected by a ground-fault circuit interrupter. Any receptacles in a bathtub or shower area must also be protected by ground-fault interrupters.

Receptacles are forbidden within 30 inches of tubs or showers.

If heat tape outlets are installed, they must be mounted within 2 feet of the cold-water pipe for which they provide protection.

Grounding

All exposed noncurrent-carrying metal parts in mobile homes must be grounded to the equipment grounding bus in the panelboard. Note that this connection is to a separate equipment grounding bus, *not* to the neutral bus. Additionally, a bonding conductor must be installed between the equipment grounding conductor bus and the chassis of the mobile home.

In addition to these specific requirements, all applicable requirements of *Article 250* must be complied with.

Wiring Methods

Except when limited by parts of this or another article, any approved method of wiring (nonmetallic-sheathed cables, armored cables, raceways, etc.) can be used in mobile homes.

RECREATIONAL VEHICLE PARKS

(Article 551)

As with the coverage of mobile homes, I will not be covering those sections of *Article 551* that apply only to requirements for the manufacture of recreational vehicles and not to the units after they have left the factory. The

focus here is on the connection of these units at the recreational vehicle (RV) park, and the wiring of the parks themselves.

When performing an installation at an RV park, it is wise to talk to the local "authority having jurisdiction" (the electrical inspector) before doing any work, as experience has proven that many electrical inspectors have very definite ideas about the wiring of RV parks. By doing this, especially if this is the first work you have done in the local area, you are likely to save yourself a bit of heartache later.

This chapter does not cover all of the definitions shown in *Section 551-2,* but you should study them carefully.

Electrical Service at RV Sites

RV parks are not required to have electrical service to every *site* (the slot into which the RV is driven and parked). In fact, they have no obligation to provide any electrical service to any of their sites at all. (For obvious reasons, most do.) But when they do provide this service, the following requirements apply:

Each RV site furnished with electricity must have at least one 15- or 20-amp, 125-volt receptacle.

At least 75 percent of all RV sites having electrical supply in an RV park must be equipped with a 30-amp, 125-volt, 2-pole, 3-wire grounding-type receptacle.

Other receptacles may also be provided at these locations, including 50-amp, 3-pole, 4-wire, 125/250-volt grounding receptacles.

All 15- and 20-amp, 125-volt receptacles must have ground-fault circuit interrupter protection.

Distribution

The secondary power distribution system in RV parks must be a single-phase, 120/240-volt, 3-wire system. No other type of electrical system is allowed by the Code, because the RVs themselves are wired for this type of system and any other type can cause a hazard. In

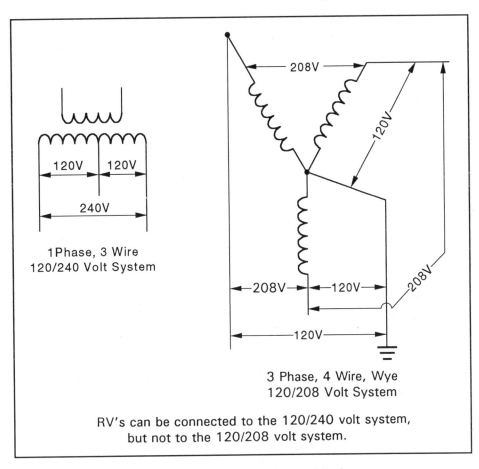

1 Phase, 3 Wire
120/240 Volt System

3 Phase, 4 Wire, Wye
120/208 Volt System

RV's can be connected to the 120/240 volt system,
but not to the 120/208 volt system.

Fig. 35-6 Voltage systems for mobile homes.

theory, it is quite possible to connect these RVs to a 120/208-volt, 3-phase, 4-wire wye system, provided it is properly done, with the RV connected to two legs of the system and the neutral. (See Figure 35-6.) However, the Code writers have chosen to make this unacceptable, since an untrained electrician could make wrong connections to such a system, causing a hazard. Therefore, only single-phase, 3-wire systems are allowed, thereby providing a somewhat more "foolproof" system.

Other requirements for an RV park distribution system are as follows:

Feeders to each site must have adequate capacity for the load being served. In no case may the ampacity be less than 30 amperes. Because of the long distances involved in RV parks, it is almost always necessary to use larger sizes of wire than the ones shown in the tables of *Article 310* to avoid excessive voltage drops, which

should not exceed 5 percent at the outlet. That is, the voltage measured at the farthest outlet cannot be more than 5 percent lower than the system's nominal voltage. For a 120-volt system, this would mean that the voltage at the farthest outlet must be at least 114 volts. If the voltage is less than this amount, larger wires must be used, which will decrease the voltage drop and so increase the voltage available at the outlet in question.

Overcurrent protection must be provided at each site and to the feeder conductors themselves.

All electrical equipment in RV parks must be grounded.

Distribution systems must be grounded at their transformer.

The neutral conductor cannot be used as an equipment ground.

No connection between a neutral and a grounding electrode conductor can be made on the load side of the service disconnecting means or the transformer distribution panelboard.

Supply Equipment

Supply equipment is the small power center to which the RV is plugged in. These units are specifically designed for this purpose, and are available at almost any electric supply house.

The requirements for the installation of these units are as follows:

Supply equipment must be located on the left side of the parked vehicle, between 8 and 10 feet away from the centerline of the *stand* (the area where the RV is expected to be parked; the slot) and, longitudinally, anywhere from the back end of the stand to a point 15 feet forward of the back of the stand. (See Figure 35-7.)

A disconnecting means or circuit supply switch must be provided in the source of supply mentioned above. (The prepackaged units you will almost certainly use have this built into them.)

All supply equipment must be accessible via a passageway at least 2 feet wide and 6-1/2 feet high cut through any trees, bushes, underbrush, etc.

Supply equipment must be mounted so that the bottom of the enclosure is at least 2 feet above ground level, and the operating means is no more than 6-1/2 feet above grade.

All equipment must be grounded by a continuous grounding conductor. Receptacles must be arranged so that their removal will not disturb the grounding system.

All outdoor equipment must be raintight.

Meter sockets without meters must be enclosed with approved blanking plates.

Overhead conductors (600 volts or less) must have a vertical clearance of 18 feet and a

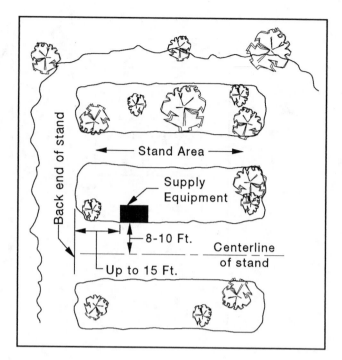

Fig. 35-7 Placement of RV supply equipment.

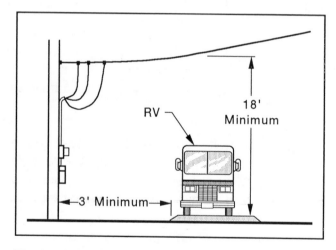

Fig. 35-8 Overhead service clearances in RV parks.

horizontal clearance of 3 feet from all areas subject to recreational vehicle movement. (See Figure 35-8.)

Underground Feeder and Branch-Circuit Conductors

All underground feeder and branch-circuit conductors must be insulated and suitable for such use. This includes aluminum, but not copper, equipment grounding conductors. Aluminum equipment grounding conductors must be insulated, but copper ones may be either insulated or bare. (In the case of corrosive soil, however, insulation is required for copper conductors also.)

All splices or taps must be made in junction boxes or by methods approved for such use, such as splicing kits suitable for the conditions.

When directly buried conductors leave a trench, they must be protected by one of the following means:

1. Rigid metal conduit.
2. Intermediate metal conduit.
3. Rigid nonmetallic conduit.
4. Electrical metallic tubing (with supplemental corrosion protection, such as bitumastic paint).
5. Other approved raceways.

The above-mentioned protection must extend from grade level down to 18 inches below grade. The generally accepted method is to take the cable from horizontal to vertical in a 90° elbow, with a vertical length of raceway extending the rest of the distance. This has proven to be the best installation technique. (See Figure 35-9.)

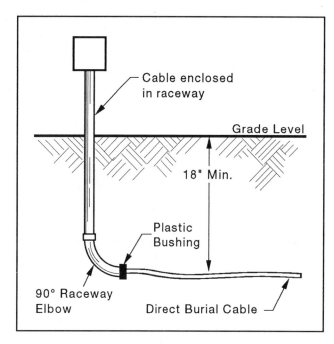

Fig. 35-9 Protection of direct burial cable.

Chapter Questions

1. When are mobile homes considered permanent residences?

2. Where are service panels for mobile homes usually installed?

3. May a supply cord to a mobile home be 20 feet long?

4. Which article of the Code covers mobile home wiring?

5. How many conductors are used for mobile home power-supply cords?

6. What special grounding is required for dryer receptacles in mobile homes?

7. Can Type NM cables be used in mobile homes?

8. What is an RV site?

9. Can 120/208-volt systems be used for RV parks?

10. What size must a passage to supply equipment be in RV parks?

CHAPTER
36

Data Processing Areas

Article 645 of the *NEC* ®is devoted completely to electronic computer and data processing installations. This article was made necessary by the vast increase in the use of computers throughout the late 1970s and 1980s. The introduction of these devices, along with their special electrical characteristics, created a lot of confusion in the electrical community, some of which still exists.

The primary confusion regarding computers concerns grounding. These questions are covered in this article (*Section 645-15* in particular) and in *Section 250-74.*

Note that this article applies only to computer rooms (also called *data processing areas*), and also that it does not cover surge suppression and other power quality concerns important to many computer installations.

The requirements in *Article 645* are as follows:

APPLICATION OF ARTICLE 645

The requirements of *Article 645* apply only to areas that have *all* of the following conditions:

1. A disconnecting means is installed, which disconnects all power in the area.
2. The area has its own dedicated heating, ventilating, and air-conditioning (HVAC) system.
3. Listed electronic and computer/data processing equipment is installed in the area.
4. The only persons permitted in these areas are those who are required for the operation of the equipment installed in them.
5. This area must be separated from other occupancies by fire-resistant floors, walls, and ceilings with protected openings. (Any openings in firewalls must be properly sealed.)
6. The construction of the building must be in accordance with building codes.

SUPPLY CIRCUITS AND CABLES

All branch circuits supplying data processing equipment must have an ampacity equal to 125 percent of the load they serve.

Data processing equipment can be connected to power circuits by any of the following means listed for the specific application:

1. Computer/data processing cables with attachment plugs.
2. Flexible cords with attachment plugs.

3. Cord-set assemblies. (If these are installed on floor surfaces, they must be protected.)

Data processing units are allowed to be interconnected with cables or cable assemblies. If these cables are run on the surface of the floor, they must be protected from physical damage.

Power cables, communications cables, connecting cables, interconnecting cables, and receptacles for data processing equipment are allowed under raised floors as long as the following conditions are met:

1. The area under the raised floor is accessible, and the floor is of suitable construction.
2. Ventilation under the floor area is used only for the data processing equipment and the data processing area.
3. All openings in the raised floors must protect cables passing through them from damage, and must minimize the amount of debris that can pass through them into the underfloor area.
4. The branch circuits to receptacles or equipment must be in one of the following:
   a. Rigid metal conduit.
   b. Intermediate metal conduit.
   c. Electrical metallic tubing.
   d. Metal wireway.
   e. Surface metal raceway that has a metal cover.
   f. Flexible metal conduit.
   g. Flexible liquidtight metal or nonmetallic conduit.
   h. Type MC, MI, or AC cable.

Power cables, communications cables, connecting cables, interconnecting cables, and associated boxes, connectors, plugs, and receptacles listed for data processing equipment and used under raised floors are *not* required to be secured in place.

Any cables that extend beyond computer rooms are subject to other sections of the Code.

POWER REQUIREMENTS

There must be one disconnecting means in the area that will disconnect all data processing equipment. There must also be a similar means to turn off all HVAC equipment in the area, and to close all fire and smoke dampers.

The disconnecting means mentioned above must be grouped together at the main exit door(s). A single disconnecting means performing both functions (disconnecting all of the data processing and HVAC equipment, and closing all fire and smoke dampers) may be used rather than two separate (though grouped together) disconnecting means. Integrated electrical systems (as described in *Article 685*) are excepted from this requirement.

Uninterruptible power supplies (UPS systems), their supply and output circuits, and battery banks must all be disconnected from the equipment they serve when the disconnecting means installed by the door is pulled. Integrated electrical systems (as described in *Article 685*) are excepted from this requirement.

GROUNDING REQUIREMENTS

All data processing equipment must either be grounded (as specified in *Article 250)* or double insulated.

Power systems derived inside data processing equipment (such as in power supplies with transformers) are not considered as separately derived systems, in regard to grounding requirements for separately derived systems *(Section 250-5[d])*.

All exposed, noncurrent-carrying metal parts of data processing equipment must be grounded.

Chapter Questions

1. Does *Article 645* apply to offices with multiple computers?

2. When must cable assemblies be protected?

3. Can rigid nonmetallic conduit be installed under raised floors in computer rooms?

4. How must smoke dampers be controlled in computer rooms?

5. Can environmental air for the building in general flow under raised computer room floors?

6. Which sections of the Code (name two) cover computer grounding?

Emergency Systems

Emergency systems are used to provide electrical service in the event of the loss of a power supply, or in some other type of emergency. The intent is that these systems will automatically provide the power and lighting required to ensure safety in emergency situations. Different applications require different levels of emergency protection. The Code covers these systems in *Articles 700, 701,* and *702,* although additional requirements for health care facilities are given in *Article 517.*

Please note that the many engineering requirements for an emergency system are not mentioned in this text—only the installation requirements. If you must design part of an emergency system, you must consult the *NEC* ®and/or a complete design handbook.

This chapter will cover the emergency system articles one at a time. (See Figure 37-1.)

EMERGENCY SYSTEMS

(*Article 700*)

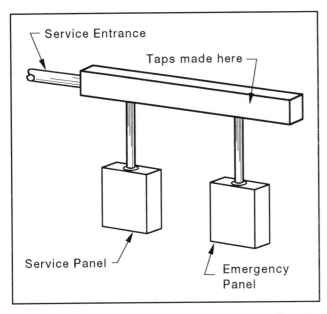

Fig. 37-1 Method of connecting emergency circuits.

Wiring of Circuits

All boxes and enclosures (of all types, including enclosures for equipment such as transfer switches, etc.) must be marked, showing that they are part of an emergency system.

All emergency system wiring must be completely independent of all other wiring systems. It may not share the same raceway, cable, box, or cabinet with other wiring systems.

The above requirement is excepted in the following circumstances:

1. In transfer switch enclosures.
2. In exit or emergency lighting fixtures (or a junction box attached to an exit or emergency fixture) that receive power from two sources.

3. Wiring from two separate emergency systems that receive their power from a common source can share raceways, cables, etc.

4. In a common junction box for a unit of equipment that contains *only* the branch circuit supplying the unit and an emergency circuit.

No appliances or lamps (except those specifically required for emergency systems) are allowed to be connected to emergency circuits.

Emergency systems must be designed and installed so that when any single lighting element fails (as when a light bulb burns out), no area will be left in total darkness.

When the normal source of lighting in an area is entirely high-intensity discharge lighting, such as high- or low-pressure sodium, metal halide, or mercury vapor (which take several minutes to get back to their original brightness after a power failure), the emergency lighting system must stay on until the normal lighting system gets back to its full brightness.

The above requirement is excepted when other means have been taken to ensure an adequate lighting level.

Branch circuits for emergency lighting must provide power from a second source when normal power is interrupted. This is done by one of the following methods:

1. An independent emergency lighting supply, to which the emergency lighting circuit will automatically transfer in the event of an outage. (The emergency lights are connected to the normal source, but are transferred [via an automatic transfer switch] to the emergency power source in case of an outage.)

2. Two or more independent sources of supply, both of which provide enough power for emergency lighting. Unless both systems are kept lighted, a means must be provided for energizing the other upon the failure of one

of the systems. Either system (or both) can be a part of the general lighting system if their circuits comply with all other emergency circuit requirements.

Branch circuits that supply emergency classified equipment must be automatically switched to an emergency power source when the normal power fails.

Switches in emergency lighting circuits must be arranged so that they can be controlled only by authorized persons. (Usually this is done by using key switches.)

The above rule is excepted if two or more single-pole switches connected in parallel are used to control the emergency lighting, and one of the switches is accessible only to authorized personnel. Also, accessible switches that can turn emergency lighting on, but not off, are allowed.

No 3-way, 4-way, or series-connected switches are allowed to be used for emergency lighting control.

All manual switches controlling emergency lighting must be accessible to the authorized persons who must operate them. In theatres or places of assembly, at least one switch must be placed in the lobby.

An emergency lighting switch may <u>never</u> be placed in a projection room, or on a stage or platform, unless it is merely one of multiple switches, and can turn the lighting on, but not off.

Exterior lights that are not required when daylight is available can be controlled by a photoelectric switch.

Branch-circuit overcurrent devices must be accessible *only* to authorized personnel.

The alternate source of power for emergency situations does *not* require ground-fault protection.

LEGALLY REQUIRED STANDBY SYSTEMS

(Article 701)

A sign must be placed at the building entrance, detailing the types and locations of legally required standby power sources.

Wiring for legally required standby systems *is* allowed to use the same raceways, cables, boxes, etc. as general wiring systems.

Branch-circuit overcurrent devices must be accessible *only* to authorized personnel.

The alternate source of power for emergency situations does *not* require ground-fault protection.

OPTIONAL STANDBY SYSTEMS

(Article 702)

A sign must be placed at the building entrance, detailing the types and locations of legally required standby power sources.

Wiring for legally required standby systems *is* allowed to use the same raceways, cables, boxes, etc. as general wiring systems.

Chapter Questions

1. Can emergency wiring share a raceway with lighting circuits?

2. Can battery lights be used instead of a generator?

3. What special requirements apply to emergency lighting switches on stages?

4. Why are emergency systems required?

5. Where can the requirement in question 1 be bypassed?

6. What do legally required standby systems need at the entrance of buildings in which they are installed?

7. Does the requirement in question 6 apply to optional standby systems also?

CHAPTER
38

Photovoltaic Systems

Photovoltaic systems use special semiconductor panels to convert solar energy into electrical energy. These systems have their own special characteristics and requirements, which are covered in *Article 690* of the *NEC* ®. Note that this article gives the general requirements for the wiring of these systems, but does not cover the design of such systems, which is a very big subject in its own right. It is not recommended that you attempt to design a photovoltaic system based solely upon your knowledge of the *NEC* ®. If you did, you might design a system that, while it would be quite safe, it probably would not work very well. You will need additional training before you will be competent in the design of these systems.

INSTALLATION REQUIREMENTS

Photovoltaic (solar power) circuits are *not* allowed to use the same raceway, cable, cable tray, outlet or junction box, or similar parts of branch circuits or feeders.

Photovoltaic circuits are permitted to be in the same box as other wiring systems if they are divided from each other by a partition.

Connections to a module or panel must be arranged so that removal of a module or panel will not disconnect a grounded conductor to another photovoltaic circuit.

Photovoltaic arrays mounted on a roof must be provided with ground-fault protection to reduce the risk of fire. This protection must be able to detect a ground fault, interrupt it, and disable the array.

In one- and two-family dwellings, photovoltaic circuits that operate at more than 150 volts to ground may not be accessible to anyone except qualified personnel.

The ratings of conductor ampacities and overcurrent devices must be at least 125 percent of the calculated current.

All photovoltaic circuits must have overcurrent protection. If a circuit has more than one source of power, overcurrent protection must be provided at each source.

Overcurrent protection must be provided for both sides of a transformer connected to a photovoltaic system. Both sides must be protected as if they were the primary.

DISCONNECTING MEANS

A means must be provided to disconnect all photovoltaic circuit conductors from all other conductors. This disconnect must meet the same general requirements as service disconnects (see *Chapter 13* of this text), but the switches do not have to be service rated, and equipment such as isolating devices, blocking diodes, and overcurrent devices are allowed on the line side of the switch.

The disconnecting means mentioned above must be a manually operated switch or breaker that is readily accessible, externally operated, and indicates whether it is open or closed; and must be rated for the load it disconnects. In cases where both sides of the disconnect can be energized when the switch is open, a warning sign must be installed that reads: "WARNING– ELECTRIC SHOCK—DO NOT TOUCH— TERMINALS ENERGIZED IN OPEN PO-SITION."

Means must also be provided to disconnect all photovoltaic equipment (power conditioners, filters, etc.) from all ungrounded conductors of any sources. If multiple sources energize the equipment, their disconnects must be grouped together and marked.

Fuses energized from both directions must be provided with a disconnecting means.

Means must be provided to disconnect all arrays or parts of arrays.

GROUNDING

Two-wire photovoltaic power sources rated over 50 volts must have one wire grounded, and 3-wire photovoltaic power sources must have their neutral conductors grounded.

The grounding connection for DC circuits can be at any one point in the circuit. (Locating the grounding point near the source of power will offer better protection.)

The equipment grounding conductor must be sized according to the following:

1. If the available short-circuit current is less than two times the overcurrent protection rating, a grounding conductor at least the same size as the supply conductors must be used.
2. If the available short-circuit current is two times or more than that of the overcurrent device, the grounding conductor must be sized according to *Table 250-95.*

All exposed metal noncurrent-carrying parts must be grounded.

WIRING

All raceway and cable methods outlined in the Code are allowed for photovoltaic installations as long as they are suitable for the area in which they are installed.

Single-conductor Type UF cables are allowed to be installed in photovoltaic source circuits. Cables exposed to sunlight must be approved for the purpose.

All interconnecting components must have the same insulation and heat tolerance characteristics as the wiring method used to provide power to the components.

All junction, outlet, and pull boxes located behind panels must be accessible, even if only by the moving of a panel.

CONNECTIONS TO SOURCES

Controls must be arranged so that a loss of power from a power conditioning unit that interacts with other power systems will cause that power unit to be disconnected from all other power sources; and will also ensure that the power conditioning unit will not be reconnected to other power systems until power from the conditioning unit is restored.

A normally interactive solar photovoltaic system is allowed to operate as a stand-alone system to supply premises wiring. (These systems do not require complex controls, and are allowed to supply power without such controls if they are not interconnected with other power systems.)

The power output from an interactive solar photovoltaic system or from a power conditioning unit must be connected to the supply side of the service disconnecting means, or to the load side as follows:

1. The connection must be made through a dedicated circuit breaker or fusible disconnect.

2. The sum of the ratings of overcurrent device currents that feed power to a conductor or bus may not be greater than the ampacity of the bus or conductor; except for systems for a dwelling unit, where this sum can be up to 120 percent of the conductor's or bus's ampacity.

3. The connection point must be on the line side of all ground-fault protective devices, except that current sources that already have ground-fault protection can be connected to the load side of other ground-fault protective devices.

4. All back-fed devices must be listed for such use.

5. In circuits that supply power to conductors or busbars, any equipment having overcurrent devices must be marked, naming all of the sources of supply to that circuit. This is not required if the circuit has only one supply source.

STORAGE BATTERIES

Since photovoltaic systems produce electricity only when the sun is shining, and since people wish to use power whether the sun is shining or not, these systems must store electricity. This requires the use of storage batteries.

The requirements for storage batteries in regard to photovoltaic systems are as follows:

All storage battery installations must conform with the requirements of *Article 480,* with the exception that interconnected battery cells are considered grounded when the photovoltaic power source is inherently protected.

In dwellings, the storage batteries must be connected so that the operating voltage is less than 50 volts, unless no parts are accessible, even during maintenance (which is frequently required).

Live parts of battery systems for dwellings must be protected to prevent accidental contact by people or objects.

A means of control must be used (whether inherently or by a separate controller) to prevent under- or overcharging the battery installations.

Chapter Questions

1. What type of device converts the sun's energy into electricity?

2. Can photovoltaic circuits and power circuits be installed in common raceways or cables?

3. Which photovoltaic arrays require ground-fault protection?

4. Which 2-wire photovoltaic sources must be grounded?

5. Where must such grounding connections be made?

6. Where in the *NEC*®are requirements for storage batteries found?

7. What special requirements apply to storage batteries operating at over 50 volts in dwellings?

8. Are back-fed photovoltaic devices ever allowed?

CHAPTER
39

Wiring Over 600 Volts

Wiring over 600 volts is covered in *Article 710* of the *NEC*®. When performing these types of installations, it is necessary to follow the manufacturer's installation instructions, not only because such high voltages must be carefully handled, but also because each installation has been individually designed from scratch and has unique characteristics.

Please note that, while wiring over 600 volts is usually called "high-voltage," this is not technically correct. Voltages between 600 and 35,000 volts are properly called "medium-voltage," and voltages over 35,000 volts are properly called "high-voltage." The requirements of this article are the basic requirements of medium-voltage wiring. (See Figure 39-1.)

WIRING METHODS

In aboveground locations, conductors can be installed as follows:

1. In rigid metal conduit.
2. In intermediate metal conduit.
3. In rigid nonmetallic conduit.
4. In cable trays.
5. As busways.
6. As cablebus.
7. In other identified raceways.
8. As open runs of metal-clad cable, when suitable for such use.
9. When accessible to qualified personnel only:
   a. As open runs of Type MV cable.
   b. As bare conductors or busbars.

Busbars can be either copper or aluminum.

Open runs of insulated wires or cables that have lead sheathing or a braided outer covering must be properly supported to avoid physical damage and electrolytic damage due to the contact between dissimilar metals.

The minimum distances between bare conductors and other conductors or grounding surfaces must be no less than the values shown in *Table 710-33*.

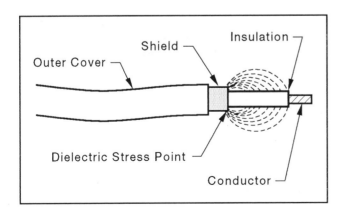

Fig. 39-1 Construction of high-voltage conductors.

UNDERGROUND CONDUCTORS

Underground conductors must be identified for the specific voltage and conditions for which they are installed.

Directly buried cables over 2000 volts must be shielded, except for nonshielded multi-conductor cables between 2001 and 5000 volts that have an overall metallic sheath or armor.

The depths of underground conductors (in raceways or as directly buried cables) must be in accordance with *Table 710-3(b)*.

Nonshielded cables must be installed in one of the following:

1. Rigid metal conduit, encased in at least 3 inches of concrete.
2. Intermediate metal conduit, encased in at least 3 inches of concrete.
3. Rigid nonmetallic conduit, encased in at least 3 inches of concrete.
4. Type MC cable, when the metal sheath is grounded by an effective grounding path (See *Section 250-51*.)
5. Lead-sheathed cable, when the lead sheath is grounded by an effective grounding path. (See *Section 250-51*.)

Conductors emerging from the ground are required to be protected by an appropriate raceway.

Conductors installed on poles must be in one of the following, from the ground to 8 feet above grade:

1. Rigid metal conduit.
2. Intermediate metal conduit.
3. Schedule 80 PVC conduit.
4. An equivalent protective means.

Conductors entering a building must be protected by an approved enclosure, from the ground line to the point of entrance.

All metallic enclosures must be grounded.

Directly buried cables may be spliced in a splice box, or without a splice box if the splice is made with an approved method suitable for the conditions encountered. Such splices must be mechanically pro-tected and watertight. If the cables are shielded, the shield must be continuous across the splice.

Backfill must not contain materials such as large or sharp rocks, which could damage directly buried conductors.

Sand or sleeves may be used to protect the cable(s).

Voltage stress reduction means must be provided at all terminations of factory-applied shielding.

Cable sheath terminating devices must be used wherever protection from moisture and physical damage is necessary.

EQUIPMENT

Equipment for high-voltage installations should be designed not only for the specific use, but also for the specific installation. There is a considerable amount of danger involved with these voltages, and every installation should be thoroughly engineered.

Pipes or ducts that require occasional maintenance, or whose malfunction would endanger the operation of electrical equipment, may not be located in the vicinity of service equipment, switchgear, or controllers. Protection from leaks, etc. must be provided.

Fire protective piping is specifically allowed.

Overcurrent protection must be provided for each conductor by an approved means.

Any doors that would give access by unauthorized persons to energized parts *must* be locked.

Control and instrument switches or push buttons must be accessible and mounted no higher than 6-1/2 feet above the floor. Operating handles that require more than 50 pounds of force to activate may not be higher than 66 inches from the floor. Operating handles for infrequently used devices may be located anywhere they can be serviced by a mobile platform.

ELECTRODE-TYPE BOILERS

Electrode-type boilers can be supplied *only* with a 3-phase, 4-wire, solidly grounded wye system; or through an isolation transformer that derives such a system.

Control circuits for electrode-type boilers must operate at 150 volts or lower, and must be taken from a grounded source. No controls are allowed in a grounded conductor.

Circuits to boilers must be rated at 100 percent of the total load.

Circuits such as mentioned above must have 3-pole ground-fault protection, which is allowed to automatically reset following an overload, but not following a ground fault.

Phase-fault protection must be in each phase.

Means must be taken for the detection of neutral and ground currents.

The neutral conductor must:

1. Be connected to the pressure vessel containing the electrodes.

2. Be insulated for no less than 600 volts.

3. Have an ampacity that is no less than that of the largest ungrounded branch-circuit conductor.

4. Be installed in the same raceway or cable tray with its circuit conductors.

5. Not be used for any other circuit.

All exposed noncurrent-carrying metal parts of electrode-type boilers and associated structures and parts must be bonded to the pressure vessel, or to the neutral conductor to which the pressure vessel is attached. The ampacity of the bonding conductor may not be less than that of the neutral conductor.

Each boiler must be equipped with a means to limit the maximum temperature and pressure by interrupting (either directly or indirectly) the current flow to the electrode. This protection must be in addition to any temperature and pressure regulation and any pressure relief or safety valves.

Chapter Questions

1. Can Schedule 40 PVC conduit be used to run wiring over 600 volts up a pole?

2. Can cables over 600 volts be spliced underground without a splice box?

3. What is the importance of *Table 710-3(b)*?

4. What is medium voltage?

5. Can wiring over 600 volts be installed in rigid nonmetallic conduit?

6. Which underground cables must be shielded?

7. What type of system must be used to supply electrode boilers?

8. How must electrode boilers be equipped to interrupt current flow?

CHAPTER
40

Communications

Article 800 of the *NEC*®covers communications circuits, such as telephone systems and outside wiring for fire and burglar alarm systems. Generally, these circuits must be separated from power circuits and grounded. In addition, all such circuits that run out of doors (even if only partially) must be provided with *circuit protectors* (surge or voltage suppressors).

The requirements are as follows:

CONDUCTORS ENTERING BUILDINGS

If communications and power conductors are supported by the same pole, or run parallel in span, the following conditions must be met:

1. Wherever possible, communications conductors should be located below power conductors.
2. Communications conductors cannot be connected to crossarms.
3. Power service drops must be separated from communications service drops by at least 12 inches.

Above roofs, communications conductors must have the following clearances:

1. Flat roofs: 8 feet.
2. Garages and other auxiliary buildings: None required.
3. Overhangs, where no more than 4 feet of communications cable will run over the area: 18 inches.
4. Where the roof slope is 4 inches rise for every 12 inches horizontally: 3 feet.

Underground communications conductors must be separated from power conductors in manholes or handholes by brick, concrete, or tile partitions.

Communications conductors should be kept at least 6 feet away from lightning protection system conductors.

CIRCUIT PROTECTION

Protectors are essentially surge arresters designed for the specific requirements of communications circuits.

Protectors must be installed on all aerial circuits not confined with a block. ("Block" here means city block.)

Protectors must also be installed on all circuits with a block that could accidentally contact power circuits over 300 volts to ground.

All protectors must be listed for the type of installation.

Metal sheaths of any communications cables must be grounded or interrupted with an insulating joint as close as practicable to the point where they enter any building (such point of entrance being the place where the communications cable emerges through an exterior wall or concrete floor slab, or from a grounded rigid or intermediate metal conduit).

Grounding conductors for communications circuits must be copper or some other corrosion-resistant material, and have insulation suitable for the area in which it is installed.

Communications grounding conductors may be no smaller than No. 14.

The grounding conductor must be run as directly as possible to the grounding electrode, and must be protected if necessary.

If the grounding conductor is protected by metal raceway, it must be bonded to the grounding conductor on both ends.

Grounding electrodes for communications grounds may be any of the following:

1. The grounding electrode of an electrical power system.
2. A grounded interior metal piping system. (Avoid gas piping systems for obvious reasons.)
3. Metal power service raceway.
4. Power service equipment enclosures.
5. A separate grounding electrode.

If the building being served has no grounding electrode system, the following can be used as a grounding electrode:

1. Any acceptable power system grounding electrode. (See *Section 250-81.*)
2. A grounded metal structure.
3. A ground rod or pipe at least 5 feet long and 1/2 inch in diameter. This rod should be driven into damp (if possible) earth, and kept separate from any lightning protection system grounds or conductors. (It is my opinion that gas pipes should not be used either.)

Connections to grounding electrodes must be made with approved means.

If the power and communications systems use separate grounding electrodes, they must be bonded together with a No. 6 copper conductor. Other electrodes may be bonded also. This is not required for mobile homes.

For mobile homes, if there is no service equipment or disconnect within 30 feet of the mobile home wall, the communications circuit must have its own grounding electrode. In this case, or if the mobile home is connected with cord and plug, the communications circuit protector must be bonded to the mobile home frame or grounding terminal with a copper conductor no smaller than No. 12.

INTERIOR COMMUNICATIONS CONDUCTORS

Communications conductors must be kept at least 2 inches away from power or Class 1 conductors, unless they are permanently separated from them or unless the power or Class 1 conductors are enclosed in one of the following:

1. Raceway.
2. Type AC, MC, UF, NM, or NMB cable, or metal-sheathed cable.

Communications cables are allowed in the same raceway, box, or cable with any of the following:

1. Class 2 and 3 remote-control, signaling, and power-limited circuits.
2. Power-limited fire protective signaling systems.
3. Conductive or nonconductive optical fiber cables.
4. Community antenna television and radio distribution systems.

Communications conductors are not allowed to be in the same raceway or fitting with power or Class 1 circuits.

Communications conductors are not allowed to be supported by raceways unless the raceway runs directly to the piece of equipment the communications circuit serves.

Openings through fire-resistant floors, walls, etc. must be sealed with an appropriate firestopping material.

Any communications cables used in plenums or environmental air-handling spaces must be listed for such use.

Communications and multipurpose cables can be installed in cable trays.

Any communications cables used in risers must be listed for such use.

Cable substitution types are shown in *Table 800-53.*

Chapter Questions

1. Why are communications conductors located below power conductors in outdoor runs?

2. How much clearance must a communications conductor have over a flat roof?

3. What is the purpose of the above requirement?

4. Is a No. 16 conductor an acceptable communications ground?

5. What is a circuit protector?

6. How far must communications wiring be kept from lightning protection systems?

7. Protectors are necessary for communications circuits that are not contained within what material?

8. What must be done if a communications ground is run in a metal conduit?

9. Can a communications system share a power system's grounding electrode?

10. How far must communications conductors be kept from Class 1 conductors?

CHAPTER
41

Special Installations

In addition to its coverage of the common wiring considerations such as grounding, wiring methods, overcurrent protection, etc., the *NEC*® covers a wide range of special installations. As I said in *Chapter 1,* the *NEC*® must provide all of the necessary requirements for every type of installation that could be insured. For this reason, there are a large number of seldom-used Code sections that cover some rather obscure topics.

In this chapter, I will explain in detail the more important—or, more correctly, the more commonly needed—sections (such as swimming pools), and simply give a listing of the unusual sections. There is simply no point in going into great detail on all of these unusual sections, such as the internal wiring of pipe organs (the topic of *Article 650*).

SWIMMING POOLS

Due to the very common usage of lighting in swimming pools, and the use of power near swimming pools, this is a very important article. Because water is usually a very good conductor of electricity, a great deal of care and special requirements must be applied to swimming pool installations.

The requirements of *Article 680,* where swimming pools are covered, are as follows (see Figures 41-1 through 41-4):

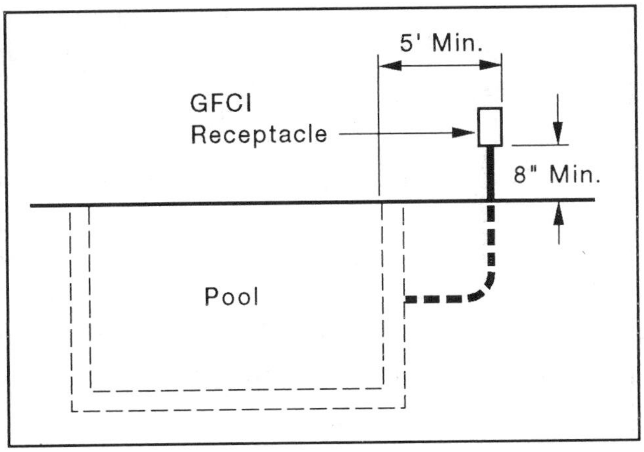

Power Supplies and Circuits

Transformers used for supplying power to underwater fixtures, along with their cases, must be identified as suitable for such installation. They must have a grounded barrier between the primary and secondary sides, and must be of the two-winding type.

Either circuit breaker- or receptacle-type ground-fault circuit interrupters (GFIs) can be used.

Conductors from the load side of GFIs cannot occupy the same enclosures or race-

Fig. 41-1 Receptacles around pools.

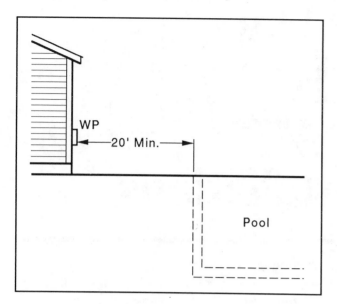

Fig. 41-2 Non-GFI receptacles around pools.

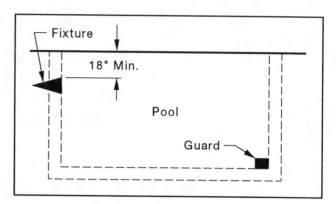

Fig. 41-3 Fixtures in pools.

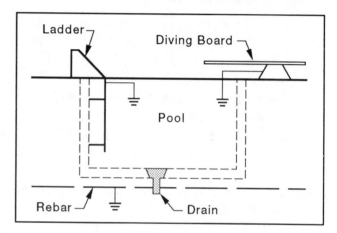

Fig. 41-4 Bonding in pools.

ways with conductors that are not GFI protected. (The panels or boxes where the GFI protected circuits originate are necessarily excepted from this requirement.)

All receptacles in a pool area must be at least 10 feet from inside pool walls, except receptacles for necessary pump motors for a permanently installed pool, which must be at least 5 feet from the inside pool wall, and must be of the locking, grounding, and GFI type.

All permanently installed pools must have at least one GFI protected receptacle between 10 and 20 feet from an inside pool wall.

All receptacles within 20 feet of the inside pool wall *must* be GFI protected.

No lighting fixtures can be installed over a pool, or within 5 feet horizontally of the inside pool wall, unless they are at least 12 feet above the highest water level, with the following exceptions:

1. If the fixtures are within 5 feet of the inside pool wall, but not over the pool, at least 5 feet above the maximum water level, and rigidly attached to the existing structure.

2. If the fixtures are totally enclosed, GFI protected, and at least 7-1/2 feet above the maximum water level.

Lighting fixtures and outlets that are between 5 and 10 feet from the inside pool wall must be GFI protected, rigidly attached to the structure, and at least 5 feet above the highest water level.

Cord-connected lighting fixtures, when installed within 16 feet of the surface of the water at any point, must have cords no more than 3 feet long, and a copper equipment grounding conductor no smaller than No. 12 AWG copper. They must also have an appropriate attachment plug.

No switches or switching devices can be installed closer than 5 feet to the inside walls of a pool, unless separated by a solid fence, wall, or other barrier.

Fixed or stationary equipment rated 20 amperes or less can be connected by the cord-and-plug method. The cord can be no more than 3 feet long, and must have a copper equipment grounding conductor no smaller than No. 12 AWG copper and an appropriate attachment plug.

Pools cannot be located under any overhead wiring (including service drops). No such wiring can be installed above a pool area, which includes any area within 10 feet horizontally of an inside pool wall; above a diving area; or above observation stands, platforms, or towers. (Utility-owned, -operated, and -maintained overhead lines can be installed above such areas. See *Figure 680-8.*) Utility-owned, -operated, and -maintained overhead communications lines can be installed above such areas if they are at least 10 feet above pools, diving areas, observation stands, platforms, or towers.

All branch circuits supplying electric pool water heaters must have an ampacity equal to 125 percent of the load they serve.

Underground wiring cannot be installed under a pool, or within 5 feet of an inside pool wall, with the following exceptions:

1. Wiring necessary to supply pool equipment can be installed in this area.
2. If space considerations demand that wiring be installed within 5 feet of the inside pool wall, the following methods can be used:
   a. Rigid metal conduit that has corrosion protection can be buried at least 6 inches below grade.
   b. Intermediate metal conduit that has corrosion protection can be buried at least 6 inches below grade.
   c. Rigid nonmetallic conduit or other approved raceways (all metal raceways require corrosion protection) can be buried at least 18 inches below grade.

Underwater Lighting

All circuits supplying underwater lighting fixtures that operate at more than 15 volts must be GFI protected.

All underwater lighting fixtures must be approved for the exact circumstances under which they are to be installed. This includes the rating and characteristics of branch circuits supplying the fixtures, as well as the mounting locations.

Bonding

Bonding is a very important part of swimming pool installations. It is this part of the system that ties together all of the grounding and life safety components. Pay close attention to the requirements, and make sure that you review them before performing (or especially designing) a swimming pool installation.

The following parts must be bonded together with a solid copper conductor (insulated or bare) no smaller than No. 8 AWG:

1. All metallic parts of the pool shell, including the reinforcing metal of pool shell, deck, and coping stones.
2. All forming shells. (These form the concrete around niche-type underwater lighting fixtures.)
3. Metal fittings in or attached to the pool structure.
4. All metal parts of electric equipment associated with the pool water system, including pump motors.
5. Metal parts of equipment associated with pool cover systems, including motors.
6. Metal-sheathed cables, metal raceways, metal piping, and all metal parts that are within 5 feet of an inside pool wall or 12 feet or less above the maximum water level or any observation stands or platforms, and that are not divided from the pool by permanent barriers.

Exceptions to the above requirements are as follows:

1. Reinforcing steel can be bonded together with steel tie wire only.
2. Isolated metal parts that do not measure more than 4 inches in any dimension and do not penetrate more than 1 inch into the pool structure do not have to be bonded.
3. Structural reinforcing steel or the walls of bolted or welded metal pool structures are allowed as a bonding grid for nonelectric parts if proper connections are made. (See *Section 250-113.*)

A bonding grid can be made of any of the following:

1. The wall of a bolted or welded pool.
2. A solid copper conductor, no smaller than No. 8 AWG, insulated, bare, or covered.
3. The reinforcing steel of a concrete pool, if the reinforcing rods are bonded together by steel tie wires or the equivalent.

Some pool water heaters rated more than 50 amperes have special bonding requirements, and the manufacturer's instructions should be followed.

Underwater Sound Equipment

All underwater audio equipment must be listed for the intended installation, and must be installed according to the manufacturer's instructions.

The following wiring methods can be used for underwater radio equipment:

1. Rigid metal conduit made of brass or other corrosion-resistant material.
2. Intermediate metal conduit made of brass or other corrosion-resistant material.
3. Rigid nonmetallic conduit.

When nonmetallic conduit is used, a No. 8 copper grounding conductor must be run

in the conduit, and terminate in the forming shell and in the junction box. In the forming shell, the No. 8 conductor must be covered with a suitable potting compound to prevent deterioration.

All electrical equipment associated with underwater audio equipment must be grounded.

Grounding Requirements

Wet-niche lighting fixtures must have a No. 12 or larger copper grounding conductor run in the conduit that feeds them.

For wet-niche fixtures wired with cords, the cords must include a separate grounding conductor that must be at least as large as the supply conductors, and may never be smaller than No. 16.

All pool associated motors must be grounded with at least a No. 12 copper insulated equipment grounding conductor.

All panelboards must have a separate equipment grounding conductor to the service panel. This conductor must be sized according to *Table 250-95,* but may never be smaller than No. 12 insulated copper.

When cords are used, their equipment grounding conductors must be connected to a fixed metal part.

OPTICAL FIBER CABLES

The inclusion of optical fiber cables in the *NEC*® is odd, in that these cables carry no electricity at all. Optical fiber cables (also called fiber optic cables) are, however, included in the *National Electrical Code*® for two primary reasons:

1. Because they are usually installed by the same persons who install electrical wiring.
2. Because optical fiber systems interact with, and depend upon, electrical and electronic systems.

As you will see in *Section 770-1,* this article does not (and really cannot) apply to installations where only optical fiber cables are being installed (that is, in places where there will be no electrical wiring installed with the optical fiber cables). Regardless of this fact, *Article 770* of the *NEC®* is the standard installation code for optical fiber cables, primarily for the reasons stated above.

The requirements of *Article 770* are as follows:

Cable Types

These cables have three basic classifications, which have to do with the type of cable jacketing and construction, rather than with the operation of the optical fibers contained in the cables. The classifications are as follows:

1. Horizontal cables.
2. Riser cables.
3. Plenum cables.

These classifications are subdivided into two divisions each:

1. Conductive.
2. Nonconductive.

Conductive cables are comprised of noncurrent-carrying metal members and are used as strength members, metal sheaths, or vapor barriers.

Nonconductive cables do not contain any such metal members.

One other type of cable is called a *hybrid cable,* as it contains not only optical fibers, but also current-carrying conductors (and possibly noncurrent-carrying conductive members also).

Installation Methods

Where optical fiber cables with noncurrent-carrying conductive members come into contact with electrical power or light conductors, the conductive member must be grounded as close as possible to the point of entrance; or, in place of grounding, an insulating joint can be installed in the conductive member, as close as possible to the point of entrance (such point of entrance being the place where the optical fiber cable emerges through an exterior wall or concrete floor slab, or from a grounded rigid or intermediate metal conduit).

Nonconductive optical fiber cables can occupy the same cable tray or raceway as electrical power, light, or Class 1 conductors operating at no more than 600 volts.

Hybrid optical fiber cables that have only power, light, or Class 1 conductors operating at no more than 600 volts in the hybrid cable may be installed in the same cable tray or raceway as electrical power, light, or Class 1 conductors operating at no more than 600 volts.

Nonconductive optical fiber cables can *not* occupy the same box, panel enclosure, etc. as the terminations of electric light, power, or Class 1 circuits, except as follows:

1. When the nonconductive optical fibers are functionally associated with the power, light, or Class 1 circuits.
2. When the nonconductive optical fibers are installed in a factory- or field-assembled control center.
3. Nonconductive optical fiber cables are allowed with circuits operating at 600 volts or less in industrial buildings when supervised only by qualified persons.
4. Hybrid optical fiber cables are allowed with circuits operating at 600 volts or less in industrial buildings when supervised only by qualified persons.

Conductive and nonconductive optical fiber cables are allowed in the same raceway, cable tray, or enclosure with any of the following:

1. Class 2 and 3 remote-control, signaling, and power-limited circuits.
2. Power-limited fire protective signaling systems.
3. Communications circuits.
4. Community antenna television and radio distribution systems.

In plenums, the following types of cables are allowed to be installed:

1. Type OFNP cable. This type of cable can also be installed in optical fiber raceways.
2. Type OFCP cable.
3. Other types of cable, installed in approved raceways. (See *Section 300-22.*)

For risers, the following types of cables are allowed to be installed:

1. Type OFNR, OFCR, OFNP, or OFCP cable.
2. Type OFN or OFC cable can be installed in one- and two-family dwellings, or in other occupancies if installed in a metal raceway or fireproof shaft, when firestops are installed at each end.

Cables in other locations can be Types OFC and OFN.

Types and Uses

Optical fiber cables must be installed according to their listings, designated as follows:

1. Type OFNP: Nonconductive optical fiber plenum cable.
2. Type OFCP: Conductive optical fiber plenum cable.
3. Type OFNR: Nonconductive optical fiber riser cable.
4. Type OFCR: Conductive optical fiber riser cable.
5. Type OFN: Nonconductive optical fiber cable.
6. Type OFC: Conductive optical fiber cable.

Types OFNP and OFCP can be used in ducts, plenums, and other spaces used for environmental air. They have low-smoke and fire-resistant characteristics.

Types OFNR and OFCR can be used in a vertical shaft, run from floor to floor. They have fire-resistant characteristics that prevent the spread of fire from floor to floor.

Types OFN and OFC can be used for general installations, but not in plenums, etc., or as risers. They are also resistant to the spread of fire.

Hybrid cables must be used exactly as listed and marked on the cable jacket.

The following types of cables can be substituted for each other:

1. For Type OFC cable: Type OFN, OFCR, OFNR, OFCR, or OFNP cable.
2. For Type OFN cable: Type OFNR or OFNP cable.
3. For Type OFCR cable: Type OFNR, OFCP, or OFNP cable.
4. For Type OFNR cable: Type OFNP cable.
5. For Type OFCP cable: Type OFNP cable.
6. For Type OFNP cable: No substitutions allowed.

THE ADVANTAGES OF OPTICAL FIBERS

A well-designed system of optical fibers as a communications medium offers four primary advantages over traditional copper wire systems:

1. **Performance.** Optical fiber cables have a much greater bandwidth operation than wire. This allows for high-volume, high-speed data transfer. Speeds can reach or exceed 90 million bits per second (90M b/s). In addition, the losses of signal strength are very small, allowing long transmissions over many miles without the need for reamplification.
2. **Electrical immunity.** Optical fiber cable is nonmetallic. Because of this, it cannot emit or pick up electromagnetic interference (EMI) or radio frequency interference (RFI), both of which are a problem with metal wires. Cross talk between cables does not exist. Additionally, optical fibers have no grounding or shorting problems.

This is also an important feature when installing communications wiring in hazardous environments. Optical fiber cabling causes no sparking or excessive heat, even when broken.

3. **Small size and low weight.** Optical fiber cable weighs approximately one-ninth as much as coaxial cable and is also physically smaller. In order to carry 24 telephone conversations, telephone companies must use two pairs of copper wire. By using optical fiber cables rather than copper, one pair of cables can carry 1,344 conversations.

4. **Security.** Electronic bugging depends on electromagnetic monitoring. Because optical fibers carry light rather than electricity, they are immune to bugging. In order for bugs to be used on optical fiber cables, the cables must be physically tapped; this is easily detectable as the signal is diminished and error rates increase.

Because of these advantages, optical fibers are increasingly being used for many types of data transmissions. It is also likely that before too long optical fibers will become the standard method of data transfer. A short list of applications follows:

1. Local area networks.
2. Telephone trunk lines.
3. Closed-circuit television.
4. Cable television.
5. Electronic news gathering.
6. Military communications.

What Optical Fiber Cable Is

Figure 41–5 shows a cross section of an optical fiber cable. As you see, it is made of three parts: a core, a cladding, and an outer coating.

The core is the part of the fiber that actually carries the light pulses used for transferring data. This core may be made of either plastic or glass. The size of the core is important, as a larger core can carry more light.

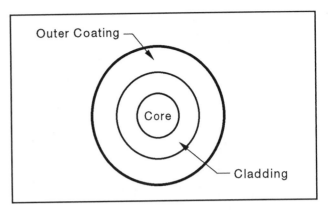

Fig. 41–5 Cross section of optical fiber cable.

The cladding sets a boundary around the fiber, so that light running into this boundary is reflected back into the cable. This keeps the light moving down the cable, keeping it from escaping.

Coatings are typically multiple layers of plastic. This is necessary to add strength to the fiber, to protect it, and to absorb shock. These coatings range between 25 and 100 microns thick (a micron is a measurement equal to one-millionth of a meter; 25 microns equal one-thousandth of an inch, the approximate thickness of one sheet of paper). Coatings can be stripped from the fiber either mechanically or chemically, depending on the type of plastic used.

CLOSED-LOOP AND PROGRAMMED POWER DISTRIBUTION

(Article 780)

This article includes requirements for a new type of electrical system designed for the National Association of Home Builders' new Smart House™ system. This system (which has advantages over existing power systems) may come into use by 1993 or 1994.

Closed-loop power systems are unique in that no power is available at the system's outlets until a proper device is plugged into it. (See Figure 41–6.) Power is sent to the various outlets through branch circuits, just as for normal wiring; but at the outlet, power is handled

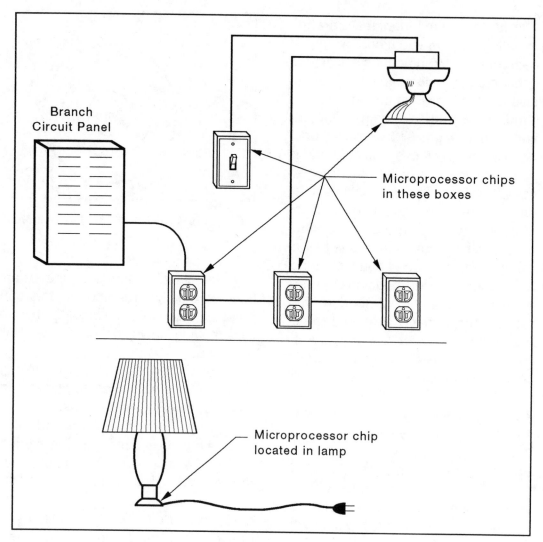

Fig. 41-6 Closed-loop wiring diagram.

quite differently. The power is received at the outlet, but not fed through to the device immediately. Each outlet has a microprocessor chip in it that communicates with a similar chip in each piece of equipment designed to be installed on the system. If the chip in the utilization equipment (lamp, radio, coffeemaker, etc.) identifies itself to the chip in the outlet, power will be sent to the equipment. If the equipment cannot properly identify itself, no power will be sent.

This system is still under development, and there are still a lot of bugs in the system. When it is fully operational, it will guarantee a high degree of safety. With this system, there can be no accidental shocks from a foreign object being stuck in the outlet. In addition, the chip in

the outlet can sense if the equipment is drawing too much power and shut it down. For instance, if a radio begins to draw more than an amp or so, the chip in the outlet will recognize that this is too much power, decide that there must be a malfunction, and shut the radio's power off.

The closed-loop power system looks fantastic on paper, but it must pass two big hurdles to become widely used: First, it has to be fully developed. At this date, it is still in the developmental stages, even though the developers have begun sales promotion. Second, this technology must become more affordable before many people will buy it. An extra $10,000 on the cost of a house is more than most people want to spend.

RADIO AND TELEVISION EQUIPMENT

(Article 810)

This article covers the requirements pertaining to radio and television transmitters, antennas, and related equipment. These types of installations are normally handled by specialists, not by general wiremen. Refer to this article if you require detailed information.

COMMUNITY ANTENNA TELEVISION AND RADIO DISTRIBUTION SYSTEMS

(Article 820)

This article covers the requirements of installations such as a central satellite receiver (a satellite dish, in the common vernacular) that feeds its signal to a group of houses or buildings. This is not uncommon in residential complexes. Refer to this article for precise requirements.

FIRE PROTECTIVE SIGNALING SYSTEMS

(Article 760)

These are the requirements that pertain to the internal wiring of fire alarm systems, and the signal and detection wiring associated with them. You can refer to the Code for all of the requirements.

INTERCONNECTED ELECTRIC POWER SOURCES

(Article 705)

This article covers power sources operated in parallel with utility company power systems. These systems are allowed to be used this way, although there are requirements for plaques by which people in the area will be warned that the parallel system is in operation.

IRRIGATION MACHINES

(Article 675)

Obviously, irrigation machines are not going to be of concern to most electricians. However, if you should need to work on one of these machines, you can find all of the applicable requirements in *Article 675* of the *NEC*®.

INDUSTRIAL MACHINERY

(Article 670)

Industrial machines can be very complex and difficult to make and maintain. For this reason, the Code gives their requirements in a separate section. Since most electricians rarely have to work on these machines, I have not covered them in this introductory book. If you need information, however, you can find it in *Article 670.*

ELECTROPLATING

(Article 669)

Electroplating is the process of covering one metal (usually some alloy of iron, copper, or brass) with a thin coating of another metal (usually zinc, tin, silver, or gold) through the use of a controlled direct current. Refer to *Article 669* for the requirements.

ELECTROLYTIC CELLS

(Article 668)

Electrolytic cells are vessels in which electrochemical reactions are caused by applying electricity. They are used in the process of refining metals such as aluminum, magnesium, cadmium, and zinc. Refer to this article for details.

PIPE ORGANS

(Article 650)

This article covers the internal wiring of organs, and is of little use to general wiremen.

Refer to the Code if you should need more information.

X-RAY EQUIPMENT

(*Article 660*)

This article covers the requirements for both X-ray machines and the wiring that feeds power to them. Since you are likely to need these requirements only once in a great while (or maybe never), I have not covered them in detail; refer to the Code.

INDUCTION HEATING EQUIPMENT

(*Article 665*)

Induction heating is an industrial process used primarily for heat treating metal parts. This article covers the electrical requirements of the machines and wiring used in this process. Refer to the Code for specifics.

SOUND-RECORDING EQUIPMENT

(*Article 640*)

Aside from a few scattered requirements and restrictions in the main body of the Code, all of the requirements covering sound-recording devices and related equipment are contained in this article. It will be important to you if you do a lot of work in recording studios, and probably of no use to you if you do not. Refer to the article for details.

ELECTRIC WELDERS

(*Article 630*)

This article covers the requirements for arc welders and similar devices, and the wiring that supplies power to them. Refer to this article for the specific requirements.

ELEVATORS

(*Article 620*)

Elevator work is a specialty in its own right. All of the electrical requirements for passenger elevators are found in this article. You will also notice many places in the Code where specific types of wiring are disallowed for use in or around elevators. Refer to the article for complete information.

MARINAS AND BOATYARDS

(*Article 555*)

Wiring in and around boats, marinas, and boatyards requires special techniques, due primarily to the water that surrounds these areas and conducts electricity. Refer to the article for more information; and definitely do not do any work in these locations without extensive study.

AGRICULTURAL BUILDINGS

(*Article 547*)

This article applies to agricultural buildings and certain parts of these buildings. The most common types of wiring are usually various types of cables. Frequently, it is also necessary to use wiring types that can handle corrosive gases. Refer to the article for complete information.

Chapter Questions

1. What parts of a swimming pool installation must be grounded?

2. How far must receptacles be kept from pool walls?

3. Must a receptacle 18 feet from a pool wall be of the GFI type?

4. Must 12-volt underwater lighting fixtures have GFI protection?

5. What are the three classifications of optical fiber cables?

6. What are conductive optical fiber cables?

7. What is a hybrid optical fiber cable?

8. What types of optical fiber cables can be installed in plenums?

9. What are the advantages of optical fiber cables?

10. What is cladding in an optical fiber cable?

11. How do closed-loop circuits use microchips?

12. Why is closed-loop wiring safer than standard wiring systems?

13. Does electroplating use AC or DC current?

Index